editors
Peter Galison
Stephen R. Graubard
Everett Mendelsohn

Science
in
Culture

T0347110

Routledge
Taylor & Francis Group

LONDON AND NEW YORK

Originally published as a special issue of *Daedalus*, Winter 1998, "Science in Culture," of the Proceedings of the American Academy of Arts and Sciences, copyright © 1998 by the American Academy of Arts and Sciences.

Published 2001 by Transaction Publishers

Published 2017 by Routledge
2 Park Square, Milton Park, Abingdon, Oxon, OX14 4RN
711 Third Avenue, New York, NY 10017, USA

Routledge is an imprint of the Taylor & Francis Group, an informa business

Library of Congress Catalog Number: 2001027724

Library of Congress Cataloging-in-Publication Data

Science in culture / Peter Galison, Stephen R. Graubard, and Everett Mendelsohn, editors.
    p. cm.
    "Originally published as a special issue of Daedalus, winter 1998"—T.p. verso.
    Includes bibliographical references and index.
    ISBN 0-7658-0673-8 (acid-free paper)
    1. Science—Social aspects. 2. Science and civilization. I. Galison, Peter Louis. II. Graubard, Stephen Richards. III. Mendelsohn, Everett.

Q175.55 .S2922 2001
306.4'5—dc21   2001027724

ISBN 13: 978-0-7658-0673-4 (pbk)

Contents

Preface

SOME TWENTY-FIVE YEARS AGO, Gerald Holton's *Thematic Origins of Scientific Thought* introduced to a wide audience his ideas about the themata. He argued that from ancient times to the modern period, there was an astonishing feature of innovative scientific work: it seemed to hold, simultaneously, deep and opposite commitments of the most fundamental sort. Continuity/discreteness, causality/randomness, unity/disunity—such antithetical dyads shed new light on the orientation and meaning of the work of Kepler, Einstein, and Bohr. Holton's own career trajectory has carried with it a thematic/anti-thematic commitment of a different sort. Over the whole course of his career he has embraced both the humanities and the sciences, looking to physics as his subject while maintaining the humanistic essay as his method. Given this background, it is entirely fitting that the explorations assembled here reflect both individually and collectively these dual roots. In fact, though each essay reads separately and informatively on its own, it is also possible to read them as a kind of map—a cartographical guide to the productive sites where Holton has turned his spade.

Our opening essay, by Holton himself, is a summing up of his long engagement with Einstein. For here are thematic commitments—to unity, for example—but expressed in the literary-philosophical context of late nineteenth-century German-speaking Central Europe. In Goethe, Boltzmann, and the neo-Kantians, Holton finds a *Kultur* that identified a class, marked an education, and presented ideals of comportment that Einstein found congenial and supportive in his scientific work.

The theme of unity expressed itself very differently as postwar American scientists reformulated the Unity of Science movement. I argue here that the wartime experience of local, pragmatic unifications in small sites like the Psycho-Acoustic or in vast laboratories such as Los Alamos or the MIT Radiation Laboratory created a new picture of scientific interaction, not reductionist but productive through shared language and techniques. This American vision of unity both appropriated the early Central European vision and yet clashed with it.

As Holton has grappled with the history of physics, he has been drawn repeatedly to the question of the scientific imagination, the title of one of his best-known books. How, Holton has asked, could one reconcile the logico-empirical character of much published science with its manifestly more volatile and exploratory formative moments? Our next two essays address this imagination, but in a thoroughly historicized form. Lorraine Daston returns the question of the scientific imagination to the Enlightenment, a period when, in their pursuit of accuracy, the sciences and arts both feared imagination in very similar ways. The split to which we have been accustomed—in which imagination was valued in the arts and loathed in the sciences—is, Daston argues, a nineteenth-century divide.

James Ackerman's chapter on Leonardo da Vinci meshes perfectly with Daston's account. For in Ackerman's da Vinci's work we see a form of imaginative intervention where it is irrelevant to draw analogies between art and science—Leonardo's critical observations of the eye are at one and the same time contributions to art, to anatomy, and to a more realistic and less geometrical perspective. Observing a recently dissected heart and combining fluid dynamics, texture, with accuracy cannot be prized apart into separate boxes for art and science.

In this da Vincian world the objective and subjective merge, as do the imaginative and the experiential.

Historians of religion Wendy Doniger and Gregory Spinner also pursue the imagination, not in the laboratory or studio, but in the bedroom and its literary-theological representations. There is startling unanimity across ancient cultures that the female imagination, impressionable as it was thought to be, could alter the form of the fetus. Doniger and Spinner suggest that such tenaciously held beliefs could be seen as imaginations of imaginations: anxious male fantasies of wandering female imaginings. Or perhaps these male imaginations were averting attention from the more obvious biological explanation for how off-springs might not resemble their fathers.

Science, culture, and the imagination intersect too in the essays by biologist Edward Wilson and physicist Steven Weinberg. Wilson takes commitment to reductionism to be a powerful driving force of science: grasping ecology in terms of organisms, organisms in terms of cells, cells in terms of biochemical processes, and biochemical processes in terms of physics. This downwards epistemic motion, he argues, has held the sciences productively together. But Wilson aims to go much further, drawing culture itself into the mix as imaging of the living brain and the examination of cross-cultural constants has begun (he argues) to suggest an evolutionary biological basis to fundamental aspects of cognition and culture. Weinberg, for his part, is also an explanation-reductionist, but not in the cultural domain. Rather, he aims to defend what string theorists have in mind when they speak of a final theory. Like Wilson, he is not defending the view that the basic sciences will yield predictions of higher-order phenomena of the type Doniger and Spinner describe. Weinberg is rather in pursuit of the end of the persistent question "why?," a chain he takes as terminating in the fundamental theories of physics and, along the way, defending a role for current physical knowledge in the assessment of the past.

If Wilson and Weinberg tackle the big questions of the unity of knowledge and worldviews from a scientific perspective, art historian Ernst Gombrich has long been after the similar territory from the perspective of art history. Here Gombrich exploits a series of sixteenth-century prints by Johannes Stradanus to track the wealth of new technologies taken from the East into Western Europe. Of course such objects as the compass, gunpowder, and paper had intrinsic significance, but for Gombrich they are equally powerful as symbols and signs of deviation from antiquity—and ultimately of progress. And it is in its abiding commitment to an "ascending line" of progress that Gombrich sees the distinctive worldview of modernity.

Holton has long been concerned with worldviews—his essay here makes it central—but there is another side to his work, a side emphasizing the nitty-gritty of scientific practice exemplified by his often-cited essay on the Millikan-Ehrenhaft dispute over the quantization of the electronic charge. Our volume would not be complete without emphasizing the texture of laboratory work—and here chemists Bretislav Friedrich and Dudley Herschbach provide us with a remarkable historical tour of the bench at the boundary of chemistry and physics. Friedrich and Herschbach track the uncertain early, sometimes halting steps that Otto Stern and Walther Gerlach took as they tried to navigate between ideas of magnetism, chemistry, atomic theory, and spectral lines. In the process of doing so, they brought space quantization from a "symbolic expression" to a part of the everyday reality of the physical laboratory. What had been a jumble of special assumptions and fragmentary insights into the atom in the planning stage became the very exemplar of textbook quantum physics.

Our final essay, by the historian of education, Patricia Albjerg Graham, addresses pedagogy head-on. Graham very helpfully periodizes educational doctrine in the United States into (roughly) the four quarters of the twentieth century. Very schematically, she suggests, these quadrants fall under the rubrics of assimilation, adjustment, access, and achievement. Apart from his teaching and textbooks, Holton's own most salient contribution to education was his participation in authoring *A Nation At Risk*, a defining document of the "achievement" phase, Graham's fourth (and our current) defining period. Graham's major historical lesson is that the rapidity of these goal-orientations has in many ways left us ambivalent about the purpose of education—and that we will need to take this shifting telos into account as we struggle to formulate educational policy in the next century.

Having all too briefly surveyed the various explorations represented in this volume, it is no doubt too much to expect a unity among them under a Sign of the Macrocosmos. We have no single banner, no binding dogma. But perhaps in these various reflections on science, art, literature, philosophy, and education, we do have one view in common: a deep and abiding respect for Gerald Holton's contribution to our understanding of science in culture.

This volume began with a symposium at the American Academy of Arts and Sciences in 1996. It is a great pleasure to thank the participants at that event, especially Harvey Brooks, Robert S. Cohen, Yehuda Elkana, Martin Klein, Robert K. Merton, and Harriet Zuckerman. Support for the symposium was graciously provided by the American Academy of Arts and Sciences and by the Department of the History of Science at Harvard University.

Peter Galison

Gerald Holton

Einstein and the Cultural Roots of Modern Science

THE ROOTS OF SCIENCE IN THE CULTURAL SOIL

THE FRUITS OF SCIENTIFIC RESEARCH are nourished by many roots, including the earlier work of other scientists. Significantly, Albert Einstein himself characterized his work as the "Maxwellian Program."[1]

But the imagination of scientists often draws also on another, quite different, "extrascientific" type of source. In Einstein's own intellectual autobiography, he asserted that reading David Hume and Ernst Mach had crucially aided in his early discoveries.[2] Such hints point to one path that historical scholarship on Einstein, to this day, has hardly explored—tracing the main *cultural roots* that may have helped shape Einstein's scientific ideas in the first place, for example, the literary or philosophic aspect of the cultural milieu in which he and many of his fellow scientists grew up.[3] To put the question more generally, as Erwin Schrödinger did in 1932: To what extent is the pursuit of science *milieubedingt*, where the word *bedingt* can have the strict connective sense of "dependent on," the more gentle and useful meaning of "being conditioned by," or, as I prefer, "to be in resonance with"? In short, the main thrust of this essay is to explore how the cultural milieu in which Einstein found himself resonated with and conditioned his science.

Gerald Holton is Mallinckrodt Professor of Physics and Professor of History of Science Emeritus at Harvard University. This essay is based on the Robert and Maurine Rothschild Distinguished Lecture in the History of Science, Harvard University, April 8, 1997.

1

There are major studies of such milieu resonances for earlier scientists: for example, the effect of the neo-Platonic philosophy on the imagination of seventeenth-century figures such as Kepler and Galileo; the theological interests that affected Newton's work; the adherence to *Naturphilosophie* that supported the discoveries of Oersted, J. R. Mayer, and Ampère; or the connection between the religious beliefs of the Puritan period and the science of the day, described in the apt metaphor that concludes Robert K. Merton's famous 1938 monograph, "The *cultural soil* of seventeenth century England was peculiarly fertile for the growth and spread of science."[4]

But there have thus far been few attempts to take up the influence of the cultural milieu on the scientific advances of twentieth-century physical scientists. The best known is that of Paul Forman, who more than two decades ago tried to interpret aspects of some scientists' presentations of quantum mechanics chiefly as their response to the sociopolitical malaise in the Weimar Republic[5]—although that work has been vigorously disputed by John Hendry, Stephen Brush, and more recently by Kraft and Kroes.[6] An example of a different sort is in an area in which Max Jammer and I have published, namely, the study of the extent to which Niels Bohr's introduction of the complementarity principle into physics was influenced by his delight in Søren Kierkegaard's philosophical writings, by his courses taken under the philosopher Harald Høffding, and also, as he claimed, by his reading of William James.[7]

But so far, there have been few such investigations in the wider, intellectual-cultural direction. I have long thought (and taught) that the full understanding of any particular scientific advance requires attention to both content and context, employing the whole orchestra of instruments, so to speak, playing out the many interacting components, without which there cannot be a full description or understanding of a case. But this is rarely done, though a middle ground exists between the extremes of internalistic study of the text alone, on one end, and constructivist externalism on the other. Moreover, in tracing the contributions of twentieth-century physical scientists themselves, the bridge from the *humanistic* parts of culture to the scientific ones—which carried much traffic in the past—has narrowed and become fragile. That is a deplorable loss, and one that deserves our attention.

The specific case of Einstein demands such attention for at least two reasons. First, it may serve as an example for studying other major twentieth-century scientists whose work has been nourished by subterranean connections to elements of the humanistic tradition. Second, it will help us resolve an intriguing paradox that has plagued scholars concerned with the source and originality of Einstein's creativity.

A PERSONAL INTERLUDE

While it is fashionable for scholars to hide assiduously the private motivations and circumstances that initiated a specific research program, on this occasion it will be useful to sketch the personal trajectory that caused me to become aware of the puzzling, paradoxical aspects of Einstein's early work.

I can fix the moment at which I was first drawn into this field of research. When the news of Einstein's death on April 18, 1955, reached our physics department, my colleagues proposed a local commemoration of Einstein's life and work. Although my own research was chiefly in experimental high-pressure physics, I had also begun to write on topics in the history of science, and so my assignment was to present how Einstein's work had been analyzed by modern historians of science. Little did I know that this suggestion would start me on a search that eventually would change profoundly my life as a scholar.

First, I discovered to my dismay that practically nothing had been done by modern historians to study seriously Einstein's *scientific* contributions—their roots, their structure, their development, their wider influence. This was in striking contrast to the volume and distinction of scholarship on the work of scientists of earlier periods, which had examined and assessed the legacy of such giants as George Sarton, Otto Neugebauer, Joseph Needham, Marjorie Nicolson, Robert Merton, Alexander Koyré, Hélène Metzger, Ludwig Fleck, and others—not to speak of *their* ancestors, such as Pierre Duhem and Ernst Mach. I seemed to be in virgin territory. Even among the many Einstein biographies, there were few serious sources.[8]

Figure 1: Albert Einstein at age nineteen. (By permission of the Einstein Archive, Hebrew University, Jerusalem.)

In truth, at the time of Einstein's death he was still deeply respected, but chiefly by way of ancestral piety and for his courageous political opposition at the time to McCarthyism, the arms race, and the Cold War. Scientists generally regarded him as having become an obstinate seeker who had wasted his last decades pursuing in vain his program of finding a unified field theory; as he told a friend, "At Princeton they regard me as the village idiot." Even his general relativity theory began to be widely taught again only after his death. In his last years, he had become a ghostly figure—a long way from the image of the vigorous young man, ready for a brilliant career (figure 1).

Today, four decades later, this perception has vastly changed. To be sure, many bubbles are bursting from the deluge of trendy journalism, whose motto in writing on major figures is well summarized in a recent essay on Herman Melville that carried the headline "Forget the Whale—the Big Question is: Did He Beat His Wife?"[9] But among the people at large Einstein's image is perhaps more ubiquitous that ever; from professional science historians, there is now an increasing flood of good scholarship on Einstein, especially since a team of researchers at Boston University has begun to publish the volumes of Einstein's *Collected Papers*, with their extraordinarily valuable editorial comments providing further stimuli for research.

None of this could have been foreseen in 1955. In retrospect, I regret not having the wit, as I was drawn into this field, to quote Marie Curie. When asked why she took up the study of radioactivity, she is said to have replied, "Because there was no bibliography." But as the historian Tetsu Hiroshige later commented, somebody had to take a "first step" in research on Einstein; eventually, it helped launch an industry analogous to the long-established ones on Newton or Darwin.[10]

That first step came in the form of a trip to the Institute for Advanced Study in Princeton to look for documents on which to base some original remarks at the memorial meeting. The key to access would be Helen Dukas, not only a trustee of Einstein's estate, but active in Einstein's household from 1928 as his secretary and later as general marshallin in the household—knowledgeable about much of his life and work, she was the untiring translator of his drafts into English and, as it turned out, endowed with an encyclopedic memory of the details of Einstein's vast correspondence that had passed through her hands.

Elsewhere I have described something of my first encounter.[11] In the bowels of Fuld Hall at the Institute was a large vault, similar to those in banks. The heavy door was partly open, and inside, illuminated dimly by a lamp on her desk, was Helen Dukas, still handling correspondence, among twenty or so file drawers that turned out to contain Einstein's scientific correspondence and manuscripts.

Once I had calmed her inborn suspicion about strangers and was allowed to have access to the files, I found myself in a state of indescribable exhilaration, in a fantastic treasure house—the kind of which most historians dream. Those documents, almost all unpublished, were arranged in a chaotic state through which only Miss Dukas knew her way with ease; they seemed to breathe

the life of the great scientist and his correspondents from all points of the compass, a rich mixture of science and philosophical speculation, of humor and dead-serious calculations.

Eventually, during two stays at the Institute, I induced Miss Dukas to help reorganize the papers into an archive suitable for scholarly research, to have a *catalogue raisonné* made, and by and by to add to the files at the Institute what she called "the more personal correspondence," which she had kept at Einstein's Mercer Street home. The whole lot, now numbering about 45,000 documents, has since been transferred by Einstein's will to the library at Hebrew University in Jerusalem. Represented in that collection are most major physicists in Europe and abroad who were living at that time, as well as authors, artists, statesmen, and the wretched of the earth, seeking help. The collection is indeed a microscope on half a century of history.

It is an amazingly diverse correspondence. Take, for example, the letters exchanged during just one of Einstein's immensely busy and creative periods (1914–18); they indicate a wide spectrum of interests among the correspondents—mostly scientists—even if gauged just by the references made to the works of major scientific, literary, and philosophical figures, including Ampère, Boltzmann, Hegel, Helmholtz, Hertz, Hume, Kant, Kirchhoff, Mach, Poincaré, and Spinoza. And one word repeatedly appeared in the correspondence—*Weltbild*, only faintly translatable as "worldpicture" or "worldview." Initially I hardly knew how important this concept, and these authors, would become in understanding Einstein's whole research program.

But to return to my mission at the time. How to proceed? In that mountain of papers at Princeton, the question of which problem I would use to start on a historical study was almost irrelevant; wherever one looked, there were exciting possibilities. For example, what role did experiments play in the genesis of the special relativity theory? Like practically everyone else, I had thought that the Michelson-Morley experiment of 1886 was the *crucial* influence that led Einstein to the relativity theory. (Indeed, I had just recently published a textbook on physics that had said so.) I had read that opinion everywhere: Robert Millikan, for example, after describing the Michelson-Morley experiment, simply concluded with the sentence, "Thus was born the special theory of relativity."[12]

But looking at samples of Einstein's correspondence, it turned out to be not so simple. One such warning occurs in his letter of February 9, 1954, to F. G. Davenport: "One can therefore understand why in my personal struggle Michelson's experiment played no role or at least no decisive role." Indeed, I later found that Einstein had repeated his stance over and over again.[13] He had typically gone his own way, relying on well-established, much older findings—experiments by Faraday, Bradley, and Fizeau—saying, "They were enough."[14] The haunting question suggests itself: what helped young Einstein make the leap when other, more established physicists could have done it so much earlier?

Another example of a key document in the files was a copy of a letter dated April 14, 1901, from Einstein to his friend and fellow student Marcel Grossmann, the existence of which was known from Seelig's biography.[15] Its eye-opening content will become clearer when we later reread that letter in the context of others in the archive. Here we need only the key sentence, in which the twenty-three-year-old beginner announces the overarching theme that would guide him through the rest of his career: "*It is a wonderful feeling to recognize the unity* [Einheitlichkeit] *of a complex of appearances, which, to direct sense experience, seem to be separate things.*"

OUTLINING THE PARADOX

These hasty first glimpses of the products of a creative mind seemed puzzling, incoherent, contradictory to me at first. They also seemed to reinforce the paradox I have mentioned before, which, in its simplest form, runs like this: It is not difficult to document that, from the start, Einstein proudly rebelled against main conventions in science as well as the social and political norms of his time. But it can be shown that at the same time he also was deeply devoted to large parts of the existing cultural canons. Was this dichotomy a hindrance, or could it possibly be a clue to understanding Einstein's uniqueness in a new way?

As to Einstein's rebelliousness, that is easily summarized in its various forms—where "rebelliousness" is a shorthand term for such traits as disobedience or insubordination to authority, a tendency to be revolutionary, obstinately nonconformist, dissident, defiant, and, in a phrase he applied to himself, "stubborn as a mule." That image of Einstein is embedded both in the public perception and throughout the literature. For example, an Einstein biography written jointly by the mathematician Banesh Hoffmann (who once worked with Einstein) and Helen Dukas herself is entitled *Albert Einstein, Creator and Rebel.*[16] Lewis Feuer, in his 1974 book *Einstein and the Generations of Science,* presented an Einstein whose whole attitude in life and science was shaped by the countercultural milieu of the throng of young revolutionaries of every sort who lived in Zurich and Bern around the turn of the century.[17] Even the *New York Times* seemed to view the confirmation of the predictions of Einstein's general relativity theory as a grave social threat. On November 16, 1919, under the title "Jazz in Scientific World," the newspaper reported at length that Charles Poor, a professor of celestial mechanics at Columbia University, thought Einstein's success showed that the spirit of unrest of that period had "invaded science," and the *Times* added its own warning: "When is space curved? When do parallel lines meet? When is a circle not a circle? When are the three angles of a triangle not equal to two right angles? Why, when Bolshevism enters the world of science, of course."[18]

But concentrating only on that aspect of Einstein overlooks an entirely different aspect of his persona, namely, Einstein as a *cultural traditionalist,* even within the limits set by his innate skepticism. If it can be proven that these opposites are combined in Einstein (as I shall show), his type of rebellion would be far from the modern image of our twentieth-century rebels in art, poetry, politics, parts of academe, or folklore—rebels who typically reject the social-political conventions of the bourgeoisie along with its cultural canon. Moreover, we shall see how Einstein's assertion of obstinate nonconformity enabled him to clear the ground ruthlessly of obstacles impeding his great scientific advance, even though the program of that advance itself ran along one of the oldest traditionalist lines. Skepticism, while necessary, was not enough to build the Temple of Isis, to use the metaphor that had long been current among German scientists.[19] Einstein made that crystal clear in a famous letter to his friend Michele Besso, who had urged him now to apply Ernst Mach's skepticism, as he had earlier, in attacking the infernal difficulties of quantum physics. Einstein replied: "You know what I think about it. [Mach's way] cannot give birth to anything living; it can only exterminate harmful vermin."[20]

We shall document Einstein's rebellious image in more detail; then examine the contradictory element; and demonstrate how the paradoxical tension between Einstein's rebellious image and his contradictory side was put to constructive use in his work. In particular, we shall examine the influence of this tension as he adopted with daring courage a set of personal presuppositions that had a history reaching back to antiquity— but for which, as he put it to Max Born, moral support came only from his own "little finger."[21] For these courageous presuppositions, on which his early success depended, he could and did draw on supporting allies— little noticed so far but far more powerful even than Einstein's little finger—i.e., ideas he had absorbed through his cultural roots, from what Merton had called, in another context, the "cultural soil" of the time.[22] In the end, we will be able to understand Einstein's program, method, and results in a new way.

AN EXCURSION INTO TERMINOLOGY

Here, a side excursion into the terminology and social stratification of Einstein's milieu is necessary. When talking about "the cultural roots of Einstein's science"—and especially today, when various definitions of "culture" are violently battling for primacy among anthropologists—a brief summary is needed of key concepts operative in the German context at the time of young Einstein's formation, in order to understand the framework within which he and his work found their place, as well as the class to which he belonged, including the aspirations of that class. The main concepts that are relevant here are *Kultur* and its companions, *Zivilisation* and *Bildung*, as well as the two composite notions of *Kulturträger* and *Bildungsbürgertum*.

The German language distinguished more sharply between *Kultur* and *Zivilisation* than did the English or French languages between their equivalents.[23] Although both *Kultur* and *Zivilisation* were generally understood in German-speaking Europe as supra-individual, collective phenomena, typically *Zivilisation* focused on the material and technological side, while *Kultur*—as first adapted in the German context by Johann Gottfried Herder—referred to the spiritual and value-related products. In extreme cases, *Zivilisation* was identified with superficial "French reason," *Kultur* with deep "German soul."[24]

At the level of the individual, the term *Bildung* (loosely translated as "intellectual formation," "self-refinement," or "education") referred to the process through which a person could acquire the attitudes and products of *Kultur*. In turn, the nation's *Kultur* as a whole was sustained—and advanced at its upper, creative level—by such *gebildete* individuals. *Bildung* thus meant much more than job-related training; it defined an ideal of human development. And a chief tool for the young to acquire *Bildung* at its best, albeit for only a small fraction of the population, was by beginning one's study in the *Gymnasium*, the neo-humanistic secondary school for ages ten to eighteen or so. The students were expected to be quite thoroughly acquainted with the great German poets and thinkers (the *Dichter und Denker*) as well as classics from other cultures, especially of antiquity.

Happily, the team now preparing Einstein's *Collected Papers* has found the curricula at Einstein's Munich schools as well as at the high school in Aarau. A quick scan of a few mandatory parts of the canon gives a good impression of how the young minds of Einstein and his cohorts were meant to be shaped. Initially there are readings from the Bible; then Latin enters at age ten, and Greek at age thirteen; Caesar's *Gallic Wars* and Ovid's *Metamorphoses* are read; then, under the supervision of his only beloved teacher, Ferdinand Ruess, poems by Uhland, Schiller, Goethe, and others; Goethe's prose poem "Hermann and Dorothea" is studied along with Xenophone's "Anabasis"; and next year, more Schiller, Herder, Cicero, Virgil. At Aarau, Einstein encounters more of the classics in German, French, and Italian; a typical entry for his course in German in 1896 reads: "History of literature from Lessing to the death of Goethe. Read

Götz von Berlichingen . . . ," and the list ends with Iphigenia and Torquato Tasso.

Such knowledge also was intended to contribute to forging a common bond between the *gebildete* individuals raised on similar *Gymnasium* curricula throughout German-speaking countries, regardless of the particular professional discipline they were later to study at the universities, whether law, medicine, the humanities, or science—a preparation for the common understanding of that class in their conversations, letters, and popular lectures, across specialties and even in their intimate personal relations.

But while the *Gymnasium* placed heavy emphasis on Latin and Greek and other aspects of "pure" *Bildung*, it had little concern for the kind of practical knowledge offered in other types of German secondary schools without such attention to classical languages, for instance, the so-called *Realschulen* (where Einstein's father Hermann and uncle Jakob, headed for electrical engineering, had received their secondary education). Needless to say, those other schools were considered, with a dose of snobbism, to be culturally less valuable; their graduates were generally not considered for university training and hence unlikely to achieve the status of *Kulturträger*.

Here it is crucial to understand a subtlety in the German concept *Kulturträger*. The term had a double meaning: both carrier and pillar of *Kultur*. On the one hand, *gebildete* individuals—chiefly the graduates of *Gymnasium* who had gone on to the universities—were seen as personally carrying or even embodying *Kultur*, living among its products, and, in the case of the most outstanding ones, advancing the *Kultur*. On the other hand, as a group they functioned also as the chief supporters (*Träger*, or "pillars") of the nation's collective project of *Kultur*. Although the term *Kulturträger* itself became generally popular only after World War I, it was a key concept earlier, as the following episode illustrates. In 1910, a bill in Prussia proposed a change in the three-tiered electoral law so that *Kulturträger* be expressly favored; they would be put into a smaller pool of voters "above the class for which their wealth would qualify them," so that their votes would count more.[25]

At the level of social stratification, most of the *Kulturträger* could be identified as belonging to what has been called the *Bildungsbürgertum* (the educated members of the bourgeoisie). The sociologist Karl Mannheim usefully distinguished two components in the modern bourgeoisie. From the beginning, he wrote, it had two kinds of social roots:"on the one hand the owners of capital, on the other those whose only capital consisted in their education."[26] In nineteenth-century Germany, the latter formed the *Bildungsbürgertum*; their social ranks were symbolized by the certificates they had attained during the process of *Bildung* and often also by a position within the hierarchies of the civil service. *Bildungsbürger* worked predominantly in professions that required university training, as physicians, lawyers, and clergy, as well as teachers and professors and other higher officials in government service.

Variants of this social stratum of the *Bildungsbürgertum* existed in many countries, but its social clout was particularly strong in nineteenth-century Germany. First, in the context of its relatively backward economy at the time, the importance of serving in the governments of the multitude of German territories large or small favored the prominence of the *Bildungsbürgertum* over the economic bourgeoisie. Second, in the absence of a nation-state and a centralized economy, German nationalism focused on *Kultur* as the basis of the nation. What held the conception of Germany together was perhaps chiefly the cultural and scholarly output of its poets and dramatists, thinkers, composers, and, eventually, its scientists. One thinks here of Goethe and Schiller, Friedrich Gottlieb Klopstock, Gotthold Ephraim Lessing, Johann Gottfried Herder, Friedrich Hölderlin and Johann Joachim Winckelmann (the prophets of Hellenism), Friedrich Schleiermacher, Friedrich Schelling, Friedrich Schlegel, Immanuel Kant, Schopenhauer, and Nietzsche, as well as Bach, Haydn, Mozart, Schubert, and Beethoven.

Thus, the academic elite among the *Kulturträger* had fundamentally a twofold mission. One was to help secure, through their scholarship, the foundation of German nationhood—though, for most of them, this also involved keeping their distance from participation in political life—and so they tended to be looked up to by those who did not, or not yet, qualify for that rank. The other was to help prepare a cadre of *gebildete* individuals, high-level functionaries who were, to adapt Fritz Ringer's terminology, "Mandarins."[27]

It is ironic that whenever Einstein, after becoming world famous, traveled abroad to lecture, an official from the local German embassy or consulate would secretly report to the foreign office in Berlin on how Einstein had behaved and how he had been received. A typical account, now available, would state that Einstein had behaved well enough, and Germany would be wise to use him to conduct what one report calls *"Kulturpropaganda."*[28] In short, he might yet be put to use as a Mandarin.

As Mannheim noted, there existed among the *Kulturträger* themselves a small group of "free-floating" (*freischwebende*) intellectuals who led marginal existences, lacked a well-defined anchor in society, and had rather critical and even rebellious inclinations. They could not or would not share the staid material comforts of the *Bildungsbürger* and disliked the whole business of "climbing up to the next rung of social existence."[29] At this point we can connect these concepts with the status and hopes of the Einstein family, asking what young Einstein's place was within the cultural-social order of the time.

The Einsteins could trace their origins in southern Germany to the seventeenth century.[30] On the male side of the family, they had largely come from the small town of Buchau, in Swabia, which in midcentury had some two thousand inhabitants, of whom a few hundred were Jews. On the maternal side, the origins were chiefly in the similarly small Swabian town of Jebenhausen. Einstein's maternal grandfather Julius Koch left Jebenhausen for Cannstadt near Stuttgart and became quite wealthy through the grain trade. Einstein's mother, Pauline, thus belonged to the bourgeoisie chiefly by virtue of capital. His father Hermann's preparation in technical school and technical trade—like that of his brother and business partner, the engineer Jakob—also did not quite qualify them as part of *Bildungsbürgertum*, and certainly not as *Kulturträger*, though one may doubt that Hermann ever gave any thought to that. But at last the family tree had sprouted, in the form of Albert Einstein, a promise to grow into that higher social region—if only the bright lad would behave as he should!

We can now reformulate the paradoxical tension of Einstein's tendency toward social-political rebelliousness and his adherence to the products of *Kultur*. Was he just one of these rootless, rebellious intellectuals, reneging on his mission as a *Kulturträger*, or did his sympathies lie with the true carriers and pillars of national culture? To make the question more graphic, imagine a scene in which Einstein first stands accused of being a free-floating intellectual intent on undermining authority, and then is defended from that charge. The testimonials offered by either side will aid in understanding better the motivations behind Einstein's behavior—and his science.

CHRONOLOGY OF A CURIOUS REBELLION

A prosecuting attorney would find it easy to establish, both by chronology as well as psychosociological profile, a portrait of Einstein as a rebellious individual throughout his life. I have no competence or deep interest in searching for the possible causes; but as to the documentable facts, many details are well known, and the pattern they form is persuasive. Einstein made his obstinacy known almost from birth, refusing to speak until about age two and a half, or, as Erik Erikson remarked, until he could begin to speak sensibly in whole sentences.[31] When Albert reached school age, his penchant for defiance took a different form. In her memoir, his sister Maja reported that in opposition to his thoroughly secular home environment, young Albert decided to become a religious Jew and accordingly "obeyed in all particulars the religious commands," including the dietary ones.[32] But after he had advanced to the Munich Luitpold Gymnasium and encountered the state-prescribed, compulsory courses on Jewish religion there, Albert's interest in Judaism came to an abrupt end. His reading in scientific books led him, as he put it in his autobiography, to the conviction that organized religious education left him "with the impression that youth is intentionally being deceived by the state through lies." He now turned to a "positively fanatic [orgy of] free thinking," having formed a "suspicion against every kind of authority"; he found solace in what he later called his "holy little book of Geometry," which was given to him as a present for self-study—a first hint of where his destiny would lead him.[33]

But as one would expect, he found school life too regimented for his taste, and he dropped out of the *Gymnasium* at fifteen and a half, surely much to the relief of some of his teachers. About a year later, he renounced his citizenship as well. When he moved for his final year of high school to Aarau, Switzerland, he arrived as a thoroughly alienated youth, having left his school, his country, and his family; he even failed in his first attempt to enroll at the Swiss Polytechnic Institute. Once Albert got into the Polytechnic he continued his "in your face" rebelliousness, to the point that when speaking to his main professor, Heinrich Friedrich Weber—on whom his career might well depend—he refused to use the obligatory title and obstinately called him just "Herr Weber." In turn, Weber did nothing to help him in his job search later.

Einstein's lifestyle at the time was distinctly bohemian.[34] He lived on the margins of bourgeois society economically, socially, and (by the standards of the day and the place) morally; he lived together with his fellow student Mileva Marić, who bore their first child before they were married in 1903. To be sure, they passionately loved each other, and as their letters show, they were of one mind in railing against the "philistine" life and conventions they saw all around them.

Even in Einstein's great paper of 1905 on relativity, one can find many touches of that self-confident defiance and seeming arrogance, not only with respect to accepted ideas in science, but also to accepted style and practice. Thus the paper contained none of the expected footnote references or credits, only a mention of his friend Michele Besso, a person who of course would be unknown among research physicists.

We shall come back to that magical first period of Einstein's brilliance. Einstein, who often characterized himself as a gypsy, at first found only temporary teaching jobs, and those tended to end abruptly and noisily; finally, after the intercession of the father of his friend Marcel Grossmann, he found refuge at the Patent Office. By 1909, he began to be sought after by universities and in 1914 accepted a call to the University in Berlin and the Prussian Academy, chiefly to gain freedom from teaching and other obligations. In fact, he managed to avoid turning out more than a single Ph.D. of his own during his lifetime.[35] As director of the Institute for Physics, his record shows that his model of leadership was to pay minimal attention to his directorial duties, even to the recruitment of new members or to drawing up regulations.[36] "Red tape," he explained, "encases the spirit like the bands of a mummy." When he first met John D. Rockefeller, Jr., the two men compared notes on how to get things done. "I put my faith in organization," Rockefeller said; "I put my faith in intuition," came Einstein's reply.[37]

When war broke out in August of 1914, ninety-three of the chief intellectuals of Germany published a manifesto with the significant title "Appeal to the World of Culture," supporting the military. Einstein, for his part, supported a pacifist counterdeclaration entitled "Appeal to the Europeans"; however, it was never published, having attracted a grand total of only four signatures. But throughout the war Einstein never made a secret of his pacifist and cosmopolitan attitude, and in an increasingly hostile Germany he took care to express publicly his support for the founding of a Jewish state in Palestine. He also made it plain that he regarded himself again as a Jew and indeed as a religious person; of course, as shown in several essays in his book *Ideas and Opinions*, his idea of religion was contrary to any religious establishment. It was a Spinozistic pantheism that he called "cosmic religion," and he put his position simply and seriously in one of his letters: "I am a deeply religious unbeliever."[38]

After his move to America when World War II broke out, the authorities kept Einstein uninformed about nuclear research. On the contrary, he was carefully monitored by the military and the FBI, which considered him a security risk. The FBI files on Einstein are voluminous; J. Edgar Hoover apparently was personally convinced that Einstein had to be watched—the physicist's whole history showed that here was a really dangerous rebel.

THE SELECTIVE REVERENCE FOR TRADITION

One could add even more weight to the side of the balance that measures Einstein's iconoclastic nature. But if now the defense attorney for the accused is given some moments for rebuttal, a counterargument might be introduced by noting that Einstein's rebelliousness was only half the story; the other half was his selective reverence for tradition. Indeed, the counsel for the defense might well urge us to consider it a hallmark of genius to tolerate and perhaps even relish what seems to us such apparent contradiction.

For there were significant limits to the offenses cited. For example, one might be more lenient about Einstein's leaving his *Gymnasium* early, since he preferred reading classics of science and literature on his own. After all, the school system was by no means beloved by all its pupils—not least because it devoted itself not only to educational goals but also to political indoctrination. Although there were variations among school systems in different parts of Germany, an official Prussian publication was typical in setting forth the plan and aims for the upper schools of 1892, when Einstein was a *Gymnasiast*. It announced, "Instruction in German is, next to that in religion and history, ethically the most significant in the organism of our higher schools. The task to be accomplished here is extraordinarily difficult and can be properly met only by those teachers who warm up the impressionable hearts of our youths for the German language, for the destiny of the German people, and the German spiritual greatness. And such teachers must be able to rely on their deeper understanding of our language and its history, while also being borne up by enthusiasm for the treasures of our literature, and being filled with patriotic spirit."[39] Clearly, *Bildung* and *Kultur* were here instrumentalized in the service of the state. To young Einstein, it smelled of militarism.

Moreover, when the public and his fellow scientists later hailed him as the great scientific revolutionary, Einstein always took pains to deny this label. He emphasized over and over again that his work was firmly embedded in the tradition of physics and had to be considered an evolution of it, rather than a revolution. He would have been appalled to know that a few years after his death a philosopher would assert that a wall of incommensurability existed between the world of Newton and the world of Einstein.

But points such as these pale in comparison to a central one: Einstein's lifelong interest in and devotion to the European literary and philosophical cultural tradition, and especially to German literary and philosophical *Kultur*. That allegiance, in which his science was clearly embedded, had been fostered early in his childhood. While the classics of music were offered in their home by his mother, Einstein's father would assemble the family in the evening around the lamplight to read aloud from works by such writers as Friedrich Schiller or Heinrich Heine.[40] The family perceived itself as participating in the movement of general *Bildung* in this way, the uplifting of mind, character, and spirit that characterized the rising portion of the *Bürgertum*. This was especially true for its Jewish segments. *Kultur* advocated and legitimized emancipation, and also provided a vehicle of social assimilation.

After all, during his scientifically most creative and intense period in Bern, Einstein formed with two young friends an "academy" for the self-study of scientific, philosophical, and literary classics. We have the list of the books they read and discussed at their meetings, which sometimes convened several times a week: Spinoza, Hume, Mach, Avenarius, Karl Pearson, Ampère, Helmholtz, Riemann, Dedekind, Clifford, Poincaré, John Stuart Mill, and Kirchhoff, as well as Sophocles and Racine, Cervantes and Dickens.[41] They would not have wanted to be *ignorant* of the cultural milieu, even if they did not necessarily agree with all they read.

To illuminate the point with but a single example: We know that Albert at the tender age of thirteen was introduced to Immanuel Kant's philosophy, starting with the *Critique of Pure Reason*, through his contacts with a regular guest at the Einstein home, Max Talmey.[42] He reread Kant's book at the age of sixteen and enrolled in a lecture course on Kant while at the Technical Institute in Zurich.[43] He wrote a lengthy book review of a philosopher's analysis of Kant, and at the Institute in Princeton his favorite topic of discussion with his friend Kurt Gödel was, again, Kant.[44] Einstein surely knew of the overwhelming influence of Kant on, for example, the late-nineteenth-century philosophers arguing against materialism.

All this, typically, did not make Einstein a Kantian at all. While sympathizing with Kantian categories—and very likely to remember that Kant had listed "Unity" as the first of his categories[45]—Einstein objected to the central point of Kant's transcendental idealism by denying the existence of the synthetic a priori, arguing: "[W]e do not conceive of the 'categories' as unalterable (conditioned by the nature of the understanding) [as Kant did], but as (in the logical sense) free conventions. They appear to be a priori only insofar as thinking without the positing of categories and of concepts in general would be as impossible as is breathing in a vacuum."[46] The essential point for him was, again, freedom, the "free play" of the individual imagination, within the empirical boundaries the world has set for us.

Thus Einstein's reverence was carefully selective, even while his outreach into the traditional cultural environment was enormous. He loved books, and they were his constant companions. A list of only those books found in the Einstein household that had been published up to 1910 includes the works of Aristophanes, Boltzmann, Ludwig Büchner, Cervantes, William Clifford, Dante, Richard Dedekind, Dickens, Dostoyevski, Frederick Hebbel, the collected works of Heine (two editions), Helmholtz, Homer, Alexander von Humboldt (both the collected works and his *Cosmos*), many books of Kant, Lessing, Mach, Nietzsche, Schopenhauer, Sophocles, Spinoza, and, for good measure, Mark Twain.[47] But what looms largest are the collected works of Johann Wolfgang von Goethe: a thirty-six-volume edition and another of twelve volumes, plus two volumes on his optics, one on the exchange of letters between Goethe and Schiller, and also a separate volume of the tragedy *Faust*, which will become a significant part of our story.

Some of those books have such early dates of publication that they may have been heirlooms; others must have been lost in the turmoil of the various migrations and separations. But this list, though only a part of the total library, indicates roughly what an aspiring member of the culture-carrying class would want to know about. And their schooling had prepared them, willing or not, to take such exemplars of higher culture seriously, not least as preparation for school examinations.

Einstein's required courses in high school were mentioned earlier; at the Polytechnic Institute, where Einstein was training to be a high-school physics teacher, he took all the obvious required science courses, including differential equations, analytical geometry, and mechanics—although what he most wanted to learn about, Maxwell's electromagnetism, he had to study on his own. In his first year, he enrolled in two additional optional courses, one on the philosophy of Kant, as noted earlier, and one entitled "Goethe, Werke and Weltanschauung." No doubt—he had been captured.

I think we now have at least an outline of the gestalt of young Einstein's complex intellectual-cultural inner life and an idea of his perception of his quite individualistic place among the *Kulturträger*—to which his whole education, both through compulsory or private reading, had carried him and where in fact he found a satisfying spiritual home.

TOWARD A VERDICT

The opposing evidences—Einstein's rebelliousness and his attention to tradition—having now been presented, is not the obvious conclusion that in Einstein we are dealing with a sort of split personality? The answer is no; we have seen two different perspectives of one coherent mental structure that uses the apparently conflicting parts to support each other.

The bonds between the apparent opposites are of three kinds. The first lies in the presence of an alternative subcurrent in the *Kultur* itself. As I have hinted, *Kultur* carried within itself a strain that we may call a "tradition of rebellion," which made it in fact potentially unstable and volatile. The anti-Enlightenment Sturm und Drang and Romantic products of the earlier period had become canonized and remained part of the tradition-bound, late nineteenth-century *Kultur*; the ideal of the active, creative, unbounded individual continued to be championed. Employing the evocative phrase Max Weber had used in a different context, such a person had to accept the plain and simple duty "to find and obey the demon that holds the fibers of one's very life"—to strive for authenticity and intensity of feeling, even heroism and sacrifice.[48] The purest expression of individuality was embodied in the genius, who led an often marginal, tormented, and, by conventional standards, failing or even demonic existence but who nonetheless saw and created things far beyond the reach of comfortable philistines.[49] Those philistines were the enemies for the Sturm und Drang authors, as they were for Einstein.

These two strains in *Kultur*, the rebellious and the traditional, often occurred in a complementary manner. Those formed by this *Kultur* were prepared to flout convention, while at the same time revering the outstanding cultural figures of all times. Though willing to dissent, they also understood themselves as loyal members of a supratemporal community of exceptional minds that existed in a universe parallel to that of the philistine masses. This mixture was not considered contradictory, although note must be taken here of what history was to record later in blood-stained letters: When these elements of rebelliousness later broke away from their stabilizing counterparts in culture, they flamed up for a time in twentieth-century Germany into the transformation and destruction of *Kultur* itself—as Einstein and so many others were to experience. But during his formative years, this complementary nature of *Kultur* still functioned, and it was precisely what Einstein needed for his work and life.

The second of the three bonds connecting those seemingly contradictory aspects of Einstein lies, unsurprisingly, in his approach to physics, both in his manner of radically clearing obstacles and in how he achieved his insights with the aid of tools from the traditional culture. Looking at his papers and letters, one can almost watch the seemingly centrifugal tendencies of Einstein's spirit being used and tamed to his service. I found the first hint in the letter he wrote in the spring of 1905 to his friend Carl Habicht.[50] In a single paragraph, Einstein poured out an accounting of major works he was then completing. First on his list is what is now known as the discovery of the quantum nature of light, as evidenced in the photoelectric effect. Another was his prediction and detailed explanation of a random, zigzag movement of small bodies in suspension that are large enough to be seen through a microscope, in which he traced the cause in exact detail to the bombardment of these visible bodies by the invisible submicroscopic chaos of molecules. (The existence of such motion, referred to as Brownian movement, was known.) And the last of the papers-in-progress he referred to was what became the original presentation of Einstein's relativity, identifying that work to Habicht only as an evolutionary act, a "modification of the teachings of space and time." To achieve that, in the published paper he casually discarded the ether, which had been preoccupying the lives of a large number of prominent physicists for more than a century, with the nonchalant remark that it was "superfluous"; dismissed the ideas of the absolutes of space, time, and simultaneity; showed that the basic differences between the two great warring camps, the electromagnetic and mechanistic worldviews, were easily dissolved into a new, relativistic one; and finally, as an afterthought, derived $E=mc^2$.

Each of these papers, completed in 1905, is a dazzling achievement, and, what is more, they always have seemed to be in three completely different fields. But I could not rid myself of the thought that behind their obvious differences something common was motivating these articles, published rapidly, one right after the other. Something was missing in that exuberant letter to Habicht.

An important lead was found at last in an unpublished letter Einstein had written to Max von Laue in January 1952, which indicates the hidden connection.[51] To put it very briefly, Einstein's study of Maxwell's theory, which had led him to the theory of relativity, had also convinced him that radiation has an atomistic (that is to say, quantum) structure, exhibiting fluctuation phenomena in the radiation pressure, and that these fluctuation should show up in the Brownian movement of a tiny suspended mirror. Thus the three separate fireworks—relativity, the quantum, and Brownian movement—had originated in a common cartridge.

Moreover, once this is understood, Einstein's approach to the problem in each of these diverse papers could be recognized as having essentially the same style and components. Unlike most other physicists of the time, Einstein did not start with a review of puzzling new experimental facts, the latest news from the laboratory, but rather by stating his dissatisfaction with what seemed to him asymmetries or other incongruities that others would dismiss as being merely aesthetic in nature. He then proposed a principle of great generality, analogous to the axioms Euclid had placed at the head of that "holy" geometry book. Then Einstein showed in each case how to remove, as one of the deduced consequences, his original dissatisfaction; at the end, briefly and in a seemingly offhand way, he proposed a few experiments that would bear out the predictions following from his theory. Once more there was only one Einstein, not three.

Most significant, the fundamental motivation behind each paper was really the very same one he had announced five years earlier in the letter to Marcel Grossmann in which he revealed what would become his chief preoccupation in science for the rest of his life: "To recognize the unity of a complex of appearances which . . . seem to be separate things." Thus, the paper on the quantum nature of light begins with a typical sentence: "There is a deep formal difference between the theoretical understanding which physicists have about gases and other ponderable bodies, and Maxwell's theory of electromagnetic processes in the so-called vacuum."[52] That is to say, energy of palpable bodies is concentrated, and not infinitely divisible; but as a light wave spreads out, its energy at a given point constantly decreases.

Why should atomicity not apply to both matter and light energy? The Brownian movement article declared that if there is chaotic motion, spontaneous fluctuation in the microcosm of classical thermodynamics, it must also show up in the macrocosm of visible bodies. And the relativity paper in effect removed the old barriers between space and time, energy and mass, electromagnetic and mechanistic worldviews. In the end, all these papers endeavored to bring together and unify apparent opposites, removing the illusory barriers between them.

THEMATIC PRESUPPOSITIONS

The longer I studied the papers and correspondence of this scientist, the more impressed I became by his courage to place his confidence, often against all available evidence, in a few fundamental guiding ideas or presuppositions, which he called "categories" in a non-Kantian sense, i.e., freely chosen. In studying other major scientists, I have repeatedly found the same courageous tendency to place one's bets early on a few nontestable but highly motivating presuppositions, which I refer to as themata. In Einstein's case, an example of themata would be simplicity, harking back to Newton's first rule of philosophy: "Nature is pleased with simplicity, and affects not the pomp of superfluous causes."[53] Einstein wrote veritable hymns to the concept of simplicity as a guide in science, and he exemplified it in his own lifestyle.[54]

Another of his thematic presuppositions was symmetry, a concept he introduced into physics in 1905, considering it basic—when most of his readers surely wrote it off as an aesthetic, optional choice. It has since become one of the fundamental ideas in modern physics. Yet another thema was his belief in strict Newtonian causality and completeness in the description of natural phenomena, which explains why Einstein could not accept as final Niels Bohr's essentially probabilistic, dice-playing universe. Einstein's utter belief in the continuum was another such thema, as in the field concepts that enchanted him from the moment he saw his first magnet compass in boyhood.

There are a few more themata to which he also clung obstinately. But beyond that, we must ask a key question: Because the themata are not a priori or innate but choosable, are those that are selected chosen at random from some infinite set of possible themata? That I do not believe. *Or are the themata so confidently held because they are reinforced by, and in resonance with, the scientist's cultural milieu?* That was the initial question here, but now it can be tested in a real case.

For that purpose, one thema that was the most important to Einstein—that of unity, unification, wholeness— will serve as the prototypical example to answer the question whether themata in science may be reinforced by the cultural milieu.[55] Einstein's dedication to the presupposition of finding unities in Nature at work is evidenced in the motivation for his three great papers of 1905. As he put it in a letter of 1916 to the astronomer Willem de Sitter, he felt always driven by "my need to generalize" (*mein Verallgemeinerungsbedürfnis*).[56] That need continued uninterrupted from his first paper on capillarity to his last ones on finding a general unified field theory that would join gravity and electromagnetism, and even provide a new interpretation of quantum phenomena—as may yet happen, although along a path different from his.[57] In between, that preoccupation had led him from the special theory to what he at first called typically the *verallgemeinerte*, the generalized theory of relativity.

That self-imposed, unquenchable desire to find unifying theories had possessed many other scientists (for example, Alexander von Humboldt, who celebrated in 1828 the "deep feeling for a unity of Nature"); however, this presupposition sometimes led Einstein astray, as had Galileo's analogous obsession with the primacy of circular motion. To be sure, some splendid science is done by researchers who seem to have no need of thematic presuppositions, as I have found in other case studies. Nor do I want to paint all German scientists as having been caught up in the dream of unity; for example, as Pauline Mazumdar's study of German immunologists showed, there were "Pluralists" among them to oppose the "Unitarians."[58]

But my subject is Einstein, and it is clear that his thematic acceptance of unity or wholeness was one of the demons that had got hold of the central fiber of his soul. He even lent his name—along with thirty-two other scholars from a great variety of fields, ranging from David Hilbert and Ernst Mach to Jacques Loeb, Sigmund Freud, Felix Klein, and Ferdinand Tönnies—to the publication, as

Aufruf!

Eine umfassende Weltanschauung auf Grund des Tatsachenstoffes vorzu-
bereiten, den die Einzelwissenschaften aufgehäuft haben, und die Ansätze dazu zu-
nächst unter den Forschern selbst zu verbreiten, ist ein immer dringenderes Bedürfnis
vor allem für die Wissenschaft geworden, dann aber auch für unsere Zeit über-
haupt, die dadurch erst erworben wird, was wir besitzen.

Doch nur durch gemeinsame Arbeit vieler kann das erreicht werden. Darum
rufen wir alle philosophisch interessierten Forscher, auf welchen wissenschaftlichen
Gebieten sie auch betätigt sein mögen, und alle Philosophen im engeren Sinne, die
zu haltbaren Lehren nur durch eindringendes Studium der Tatsachen der Erfahrung
selbst zu gelangen hoffen, zum Beitritt zu einer Gesellschaft für positivistische
Philosophie auf. Sie soll den Zweck haben, alle Wissenschaften untereinander in
lebendige Verbindung zu setzen, überall die vereinheitlichenden Begriffe zu ent-
wickeln und so zu einer widerspruchsfreien Gesamtauffassung vorzudringen.

Um nähere Auskunft wende man sich an den mitunterzeichneten Herrn
Dozent M. H. Baege, Friedrichshagen b. Berlin, Waldowstraße 23.

E. Diedgen, Fabrikbesitzer u. philos. Schriftsteller Bensheim.	**Prof. Dr. Einstein,** Prag.	**Prof. Dr. Forel** Yvorne.
Prof. Dr. Föppl, München.	**Prof. Dr. S. Freud,** Wien.	**Prof. Dr. Heim,** Geh. Hofrat, Dresden.
Prof. Dr. Hilbert, Geh. Reg.-Rat, Göttingen.	**Prof. Dr. Jensen,** Göttingen.	**Prof. Dr. Jerusalem,** Wien.
Prof. Dr. Kammerer, Geh. Reg.-Rat, Charlottenburg.	**Prof. Dr. B. Kern,** Obergeneralarzt u. Inspekteur der II. Sanitäts-Inspektion, Berlin.	**Prof. Dr. F. Klein,** Geh. Reg.-Rat, Göttingen.
Prof. Dr. Lamprecht, Geh. Hofrat, Leipzig.	**Prof. Dr. v. Liszt,** Geh. Justizrat, Berlin.	**Prof. Dr. Loeb,** Rockefeller-Institute, New-York.
Prof. Dr. E. Mach, Hofrat, Wien.	**Prof. Dr. G. E. Müller,** Geh. Reg.-Rat, Göttingen.	**Dr. Müller-Lyer,** München.
Josef Popper, Ingenieur, Wien.	**Prof. Dr. Potonié,** Königl. Landesgeologe, Berlin.	**Prof. Dr. Rhumbler,** Hann.-Münden.
Prof. Dr. Ribbert, Geh. Medizinalrat, Bonn.	**Prof. Dr. Roux,** Geh. Medizinalrat, Halle a. S.	**Prof. Dr. J. C. S. Schiller,** Corpus Christi College. Oxford.
Prof. Dr. Schuppe, Geh. Reg.-Rat, Breslau.	**Prof. Dr. Ritter v. Seeliger,** München.	**Prof. Dr. Tönnies,** Kiel.
Prof. Dr. Verworn, Bonn.	**Prof. Dr. Wernicke,** Oberrealschuldirektor u. Privat-Dozent, Braunschweig.	**Prof. Dr. Wiener,** Geh. Hofrat, Leipzig.
	Prof. Dr. Th. Ziehen, Geh. Medizinalrat, Wiesbaden.	
M. H. Baege, Dozent d. Freien Hochschule Berlin Friedrichshagen.		**Prof. Dr. Petzoldt,** Oberlehrer u. Priv.-Dozent, Spandau.

Figure 2: Appeal for the formation of the Gesellschaft für positivistische Philosophie.
(Courtesy of Wilhelm-Ostwald-Archiv, Deutsche Akademie der Wissen-schaften
zu Berlin.)

early as 1912, of a public manifesto (*Aufruf*, figure 2)
calling for the establishment of a new society aiming to
develop, across *all* branches of scholarship, one set of unifying ideas and unitary conceptions. As the *Aufruf* put it in
its second paragraph, the new Society's aim would be "to join
all fields of learning [*alle Wissenschaften*] together in an organic

association, to develop everywhere the unifying ideas, and thus to advance to a non-contradictory comprehensive conception."[59]

Yet if it was allegiance to a few themata that supported Einstein in launching into uncharted territory, often with the barest encouragement from the phenomena, *what provided the courage to adopt these themata*, and to stick with them through thick and thin? This is where the various strands we have pursued will converge, where we make closest contact with the "cultural soil" that helped to feed his scientific imagination, for one can show the resonance between Einstein's thematic belief in unity in *science* and the belief in the primacy of unity contained in certain *literary works* to which he had allegiance. While here I can demonstrate the case for only one of his themata, and for one set of major literary works, the case made is more general and applies not only to this particular scientist.

THE CULTURAL ROOTS OF UNITY—A POET POINTS THE WAY

So far, we have noted that Einstein drew on the work of other scientists, on the tools of his trade that he assembled during his education—so joyfully by himself, less so in his schooling. We have discussed his personal attitude as a *gebildete* individual, who refused to be a mere functionary of the state and kept his freedom of imagination and destiny. Other useful suggestions for pieces of the puzzle have also been proposed, for example, the interesting point made by Robert Schulmann and Jürgen Renn that Einstein's reading in popular scientific books as a boy consisted largely of ones that did not dwell on details but instead provided an overview of science as a coherent corpus of understanding, and that this experience predisposed him early to fasten upon the big questions rather than the small pieces.[60]

All this was necessary; but it was not enough. His wide reading in humanistic works beyond science—where the *Bildung* during his formative years was to lead to continued self-refinement through study of the "best works," analogous to Matthew Arnold's concept of culture—hinted at what else was needed to understand his particular genius.[61] From the list of icons of high culture at the time who greatly impressed Einstein, I must focus on just one author, indeed one who, with Friedrich Schiller, was among the most universally revered: Johann Wolfgang von Goethe.[62] Since Goethe is today certainly not on everyone's mind, I will attempt to convey in a few words his unimaginable influence at the time, not merely on educated Germans in general, but on German scientists in particular.[63]

There are two major parts to that influence. One was the fact that Goethe was arguably Germany's most accomplished and productive poet. He began his long and fruitful career when, as we noted, Germany was not a modern state. Indeed, in many ways it was backward compared with Britain and France; it was politically impotent, a motley assembly of about three hundred fragments, large and small, within the dying Holy Roman Empire. In 1775, when the twenty-six-year-old Goethe arrived in Weimar, it was still an impoverished duchy, and his own youthful presence there was possibly one of its biggest assets. His skill, intelligence, and humanity had begun to show itself even in his first, fiery works that were still linked to the Sturm und Drang tradition, for example, the irreverent revolutionary drama *Götz von Berlichingen*, written at age twenty-four, and the romantic novel *The Sorrows of Young Werter*, written one year later. The *Götz* drama was based on a legendary early sixteenth-century German knight, a bold and impudent adventurer who made it known to all, in strong language, that he was beholden to no one but God, Kaiser Maximilian, and his own independent self. (I find it delightful that during Einstein's final *Matura* examination, his essay in the subject of German was on *Götz*, the very embodiment of the independent individual spirit.[64])

Goethe, too, was a complex of apparent opposites. In his early works he had established himself as the foremost German spokesman for the Sturm und Drang movement, the forerunner of the Romantic revolt, while still adhering to Enlightenment ideas (one of those contradictions when viewed from our level below). And he was still in his twenties when he began work on the first part of his *Faust*, the tragedy into which he poured his superb poetic skills and all the varied and mutually antagonistic aspects of his maturing soul. It was, like much of his writings, part of a "great confession," but it had an especially strong grip on the German imagination, on the upward-striving bourgeoisie as well as the elite; the nearest analog that comes to mind is the indelible impression of Dante's epic on intellectuals in Italy. As G. H. Lewes remarked, the Faust tragedy "has every element: wit, pathos, wisdom, farce, mystery, melody, reverence, doubt, magic, and irony."[65]

In his early period, Goethe himself, like his Faust, accepted the dictum "To live, not to learn." But this rebellion took a special form as he matured, similar to Einstein's own. Goethe's "central tenet" was the belief in individuality or individualism: one was a free person, defying some of the social conventions but at the same time revering the geniuses of history and legend, which for him (according to Goethe's biographers) included the original Dr. Johann Faustus of the sixteenth century, Prometheus, Spinoza, Mohammed, Caesar, and the original knight Götz von Berlichingen.[66] Like Spinoza, Goethe saw God and Nature as two aspects of the same basic reality, and in that belief, too, he shared the spirit of Einstein and other scientists. Among German *Kulturträger*, Goethe became a fascinating and inexhaustible part of their imaginative lives.[67]

I will return to that point in a moment. But it must be noted that a second aspect of Goethe's power was his position as a serious and productive scientist on certain topics, such as the investigation of the subjective impression of color; the discovery, in his first scientific paper, of the presence of an intermaxilliary bone in man; his early version of what Ernst Haeckel later called an evolutionary mechanism; his concept of the metamorphosis of plants, and other such matters. Thus Goethe has an honored place even in the modern *Dictionary of Scientific Biography*, and despite the huge controversy about others of his contributions, especially on the theory of colors (the *Zur Farbenlehre* of 1810), his scientific activities—totaling fourteen volumes of the Weimar edition of his collected works—added to his standing as a figure representing the best of culture in all its dimensions.

To be sure, Goethe's science was chiefly that of the poet-philosopher. For example, one early "scientific" essay, entitled "Study after Spinoza," begins with the sentence "The concept of being and of completeness is one and the same"; from this, Goethe goes on to ponder the meaning of the infinite.[68] But significantly, the main point of that work was to argue for the primacy of unity in scientific thinking, and for the wholeness "in every living being." The sorry and misguided war he waged for over four decades against Newton's ideas, especially on color theory, must be understood in terms of Goethe's philosophical and poetic beliefs. For example, the quantification and subdivision of natural phenomena, he thought, missed the whole point of the organic unity of man and nature in the explanation of phenomena, particularly for what he regarded as qualities, such as colors. This is a prominent aspect of much of Goethe's whole corpus: the theme of unity, wholeness, the interconnection of all parts of nature. Those are main conceptions that informed both his science and his epics. As one of his commentators has mentioned, "The nature of the entire cycle [is this]: unity in duality."[69] It pervaded even his belief in the existence of an original, archetypal plant (*Urpflanze*), an archetypal man, and so on—all part of what has been called the Ionian Fallacy, looking for one overarching explanation of the diversity of phenomena.[70] Even at age eighty-one, two years before his death, he was immensely excited by news that in France, the biologist St. Hilaire had associated himself with the concept of unity at the base of biology, and he exclaimed:

> What is all intercourse with Nature, if we merely occupy our-selves with individual material parts, and do not feel the breath of the spirit which prescribes to every part its direction, and orders or sanctions every deviation by means of an inherent law! I have exerted myself in this great question for fifty years. At first I was alone, then I found support, and now at last, to my great joy, I am surpassed by congenial minds.[71]

Much has been written about the interest among scientists in various aspects of Goethe's work, and not only in Germany. A list of such scientists would contain names such as Johann Bernhard Stallo, Wilhelm Ostwald, the physiologist Arnold Adolphe Berthold, the neurophysiologists Rudolf Magnus and Emile du Bois-Reymond, the botanist Gottlieb Haberlandt, the physical chemist Gustav Tammann, the bacteriologist Robert Koch, the psychologist Georg Elias Müller, and the English scientist William Henry Fox Talbot. A curious case is that of Nicola Tesla, who, although not German by descent, was so caught up in the German style of *Bildung* that he claimed, and sometimes demonstrated, that he knew the whole of Goethe's *Faust* by heart—all 12,110 lines.[72]

Of course not everyone shared Tesla's enthusiasm. Many a scientist had to give lip service to Goethe's dominance while actually fighting for a down-to-earth, pragmatic, properly experimental style of thought. But wherever these readers turned, from their school days on, they, like Einstein, were likely to encounter Goethe and so were liable to absorb and sympathize with that central point in Goethe's work, the longing for unity, for wholeness, for the interconnectivity of all parts of nature. As Walter Moore put it in his biography of Erwin Schrödinger, "All German-speaking youth [were] imbued with the spirit of Goethe. . . . They have absorbed in their youth Goethe's feeling for the unity of Nature."[73] Fragments of Goethe's poetry could be encountered routinely, not only in the popular lectures of other *Kulturträger* or in the exhortations of politicians, but even in the lectures and textbooks on science itself, in the writings of physicists such as Hermann von Helmholtz, Erwin Schrödinger, Wilhelm Wien, and Max Born. Thus Arnold Sommerfeld, in the third volume of his *Lectures on Theoretical Physics*, sends his readers on the general relativity theory off with a quotation from *Faust*, part II.[74]

My favorite example of that ubiquity occurs on two pages of a textbook by one of Einstein's own scientific predecessors, one whom in 1900 he had called "quite magnificent."[75] Ludwig Boltzmann's *Vorlesungen über Maxwells Theorie der Elektricität und des Lichtes* was published in two parts (1891 and 1893), each preceded by a short epigraph. Boltzmann could count on every German reader to recognize the origin of the lines he quoted there, for they referred to the early pages of Goethe's *Faust* tragedy. My free translation of the first passage is: "That I may no longer, with sour labor, have to teach others that which I do not know myself"; Boltzmann does not even have to add the next, most celebrated and programmatic lines of *Faust*: "*and that I may perceive what holds the world together in its innermost.*"

Boltzmann's second epigraph refers to the passage given in italics below where Faust has just opened the book of Nostradamus, seeking even there a guide to the force that holds the world together; he gazes at the wondrous "Sign of the Macrocosm" and exclaims:

> Ha! as I gaze what rapture suddenly
> begins to flow through all my senses! ...
> *Did some god inscribe these signs*
> *that quell my inner turmoil,*
> *fill my poor heart with joy,*
> *and with mysterious force unveil*
> *the natural powers all about me?*
> Am I a god? I see so clearly now!
> In these lines' perfection I behold
> creative nature spread out before my soul. ...
> How all things interweave as one
> and work and live each in the other.[76]

By referring to the God-like signs Boltzmann meant of course to indicate Maxwell's equations, the summary of Maxwell's synthesis of electricity, magnetism, and optics. The equations relating the electric and magnetic field terms are indeed stunningly beautiful in their simplicity, scope, and symmetry, particularly when written in modern form:[77]

$$\text{curl } \mathbf{E} = -\frac{1}{c}\frac{\delta \mathbf{B}}{\delta t} \qquad \text{div } \mathbf{E} = 0$$

$$\text{curl } \mathbf{B} = \frac{1}{c}\frac{\delta \mathbf{E}}{\delta t} \qquad \text{div } \mathbf{B} = 0$$

But back to the enchanted Boltzmann. It is quite significant that in both epigraphs Boltzmann's version of Goethe's lines are in fact just a bit wrong.[78] He too was no doubt quoting from memory, going back to school days. Used constantly, such verses tend to be taken for granted and get fuzzy at the edges. Boltzmann's errors are really one sign that Goethe's lines have become part of common culture.

But never mind. We must dig a bit deeper to see why such literary allusions were so meaningful to the scientific reader. Consider the context of those lines, near the beginning of the first part of the *Faust* tragedy. Having painfully worked his way through every major specialty, Faust's thirst for knowledge at its

Figure 3: Rembrandt's etching, called "Dr. Faustus" (detail). From L. Münz, ed., *Rembrandt's Etchings* (London: Phaidon Press, 1972).

deepest level had not been satisfied by these separate (let us say, reductionist) studies—any more than were the signers of the 1912 *Appeal* for unity throughout all sciences and scholarship. Even if he has to turn to the realm of the magical, Faust must discover the secret of the world's coherence. Nostradamus's book offers him the blinding revelation in terms of the Sign of the Macrocosmos, that ancient symbol of the connection between the part and the whole, man and nature (figure 4). This is why Boltzmann connects the passage to Maxwell's equations, which express the synthesis of large parts of physics.[79]

The main point here is the strong resonance between the Goethean or Faustian drive toward a unified fundamental understanding of nature, symbolized by the Sign of the Macrocosmos, and that of the analogous ambition of Boltzmannian scientists and their pupils: the search for one single, totally coherent worldpicture, a *Weltbild* encompassing all phenomena. Physical science, too, yearned to progress by the discovery of ever fewer, ever more encompassing fundamental concepts and laws, so that one might achieve at last what Max Planck called, in the title of his 1908 essay, "Die Einheit des physikalischen Weltbildes."[80] Indeed, some physical scientists still work toward the day when one single equation, one world equation, will be found that will subsume all the diversity of physical phenomena. Then the Sign of the Macrocosmos will indeed stand before our gaze.

Einstein, starting with his very first publication in 1901 on capillarity, was committed to an early stage of such a Faustian plan. In that paper he tried to remove a duality between Newtonian gravitation, which directs the motion of macroscopic objects downward, and capillary action, which drives the molecules of the submicroscopic world of the liquid upward. In its way this was also a search for the commonality between the macrocosm of observable gravitation and the microcosm of molecular motions. Here was a case where, he thought, apparently opposite phenomena could be brought into a common vision. Even though Einstein later dismissed the physics he had used in that first paper as juvenilia, he never turned his back on the inherent goal.

Perhaps its most eloquent expression appears in his address of 1918, "Principles of Research," given in honor of Max Planck.[81] There he solemnly states that into the shaping of a coherent worldview every serious artist, philosopher, or scientist, each in his own way, "places the center of gravity of his emotional life." Einstein called that search for a worldpicture *"the supreme task"* of the physicist—the task "to arrive at those universal elementary laws from which the cosmos can be built up. . . ."

The intensity of the impulse toward a unified *Weltbild*, so typical for many German scientists of the time—even while specialization was rising all around them—was not confined to them. David Cassidy has noted that

the "unifying spirit," as it was called, pervaded much of central European thought at the turn of the century. German idealism, neo-Romanticism, and historicism, stretching from Immanuel Kant and Georg Wilhelm Hegel to Benedetto Croce and Wilhelm Dilthey, each pointed to some sort of transcendent higher unity, the existence of permanent ideas or forces that supersede or underlie the transient, ephemeral world of natural phenomena, practical applications, and the daily struggle of human existence. The scholar, the artist, the poet, the theoretical physicist all strove to grasp that higher reality, a reality that because of its permanence and transcendence must reveal ultimate "truth" and, hence, serve as a unifying basis for comprehending, for reacting to, the broader world of existence in its many manifestations.[82]

I can only add in closing that movements as different as turn-of-the-century Monism, and later the "Unity of Science" movement, were closely related to this set of aims and ideas. And as Anne Harrington has shown in her recent book, the "'holistic' biological impulse" in early nineteenth-century Germany later flourished with the assistance of our poet. As she put it, "Goethe's resulting aesthetic-teleological vision of living nature would subsequently function as one of the later generations' recurrent answers to the question of what it 'meant' to be a holistic scientist in the grand German style."[83]

At this end of Einstein's century, many excellent scientists and some philosophers are ready to settle for a hierarchical or "disunited" science rather than participate in the pursuit of overarching unities.[84] To them, the self-imposed task of those earlier culture-carriers in search of grand unifications appears perhaps overreaching, and even discussing it as a historic fact may be written off as nostalgia. Moreover, Ernst Mayr and E. O. Wilson have long insisted that for modern evolutionary biologists and naturalists the chief guiding concept should be diversity rather than unity. Perhaps Henry Adams was right when he wrote that after the nineteenth century the course of all history will be away from unity, and toward multiplicity and fragmentation.[85]

Yet the fundamental motivation of Einstein's program has helped to keep alive the modern idea of a search for a physical theory that will encompass all phenomena, from gravitation through nuclear science (a path that Einstein had not explored).

The ascent to that Mt. Everest is now taking various forms among different camps, along different routes. The physics journals and even the daily papers are witnesses; and the International School of Physics has announced a physics seminar at Lake Como with the title, in part, "A Probe of Nature's Grand Design." I have little doubt that hovering there above the audience will be a throng of kindred ghosts, including Kant, Maxwell, Boltzmann, and Einstein, and of course, among the poets, Goethe, with Faust himself next to him; and, way in back, the Greek philosopher Thales of Miletus in Ionia, who twenty-six hundred years ago had launched that Ionian dream, the thema that all things are made of *one* essence. All of those forebears had tilled and seeded the cultural soil of their time and, in turn, in their different ways, had been nourished and reinforced by it.

ASHES INTO THE WINDS

When death approached to claim Einstein in April of 1955, his last acts were still fully in character. He remained strong-willed to the end, obstinately adhering to his ways. He had recently signed a manifesto with Bertrand Russell and others, intending to bring together the international community of scientists as a unifying counterweight against the divisive, national ambitions then rampant during the arms race. For seven years, Einstein had known that a growing intestinal aneurysm of his aorta might rupture at any time, but he had refused any major operation when it still might have averted the threat. He explained his uncomplaining state of mind to his stepdaughter Margot by saying simply, "I have done my thing here." At about one o'clock in the morning, as the aneurysm burst, he suddenly spoke once more, but the night nurse did not understand German.

Einstein's requests concerning his last rest also bore all the marks of his lifelong struggle for simplicity and against ordinary convention. There was to be no funeral—only a few family members and friends gathering at the crematorium. No speeches, no flowers, not even music. No gravestone. But as Einstein's ashes were dispersed into the winds, an old friend and fellow émigré felt moved to recite a few verses of poetry, ending with these lines:

He gleams like some departing meteor bright,
Combining, with his own, eternal light.

As it happened, the poem had been written a century and a half earlier, by the grief-stricken Goethe on the occasion of the death of his friend Friedrich Schiller. A great circle had closed. Symbolically, Einstein's lifelong comrades had helped him, once more, to move across those illusory divisions between space, time, and cultures.

ACKNOWLEDGMENTS

In addition to thanking Robert and Maurine Rothschild for their support to the History of Science department, I also wish to acknowledge several colleagues whom I have consulted on aspects of this work, including Gordon Craig, Frederick Gregory, Roald Hoffmann, Robert Schulmann, S. S. Schweber—none of whom are responsible for possible errors—and above all, Gerhard Sonnert, who provided essential and dedicated help throughout. I am grateful to the Andrew W. Mellon Foundation for support of a research project of which this essay is part.

ENDNOTES

[1]It is symbolic that among the framed portraits he kept in his Princeton home there were only three scientists, each of whom pursued a great synthesis in physics—Newton, Faraday, and Maxwell.

[2]Albert Einstein, "Autobiographical Notes," in *Albert Einstein: Philosopher-Scientist*, ed. Paul Arthur Schilpp (Evanston, Ill.: Library of Living Philosophers, 1949), 53.

[3]Looking in the opposite direction, e.g., how cultural elements later were affected *by* relativity theory rather than initially helping to shape it, we know how certain of Einstein's publications were interpreted to affect the culture of his and our time, misguided though most of these attempts have been—as Steven Weinberg recently warned—such as the transfer of relativity concepts into anthropology, ethics, religion, literature, and to the so-called relativism haunting other fields. Einstein himself was perturbed by popular misunderstanding of his theory. He would have preferred if his theory—which Max Planck and Max Abraham, not Einstein himself, had named in 1906 the "theory of relativity"—had become known as the "theory of invariance" instead. Einstein, letter to E. Zschimmer, 30 September 1921; cf. Gerald Holton, *Einstein, History, and Other Passions: The Rebellion Against Science at the End of the Twentieth Century* (Reading, Mass.: Addison-Wesley,

38   Gerald Holton

1996), 131–132. See also Steven Weinberg, "Sokal's Hoax," *New York Review of Books*, 8 August 1996, 11–15.

[4]Robert K. Merton, *Science, Technology and Society in Seventeenth Century England* (New York: H. Fertig, 1970; first published 1938), 238.

[5]Paul Forman, "Weimar Culture, Causality, and Quantum Theory, 1918–1927: Adaption by German Physicists and Mathematicians to a Hostile Intellectual Environment," *Historical Studies in the Physical Sciences* 3 (1971): 1–115.

[6]John Hendry, "Weimar Culture and Quantum Causality," *History of Science* 18 (1980): 155–180; Stephen G. Brush, "The Chimerical Cat: Philosophy of Quantum Mechanics in Historical Perspective," *Social Studies of Science* 10 (1980), 393–447; P. Kraft and P. Kroes, "Adaption of Scientific Knowledge to an Intellectual Environment: Paul Forman's 'Weimar Culture, Causality, and Quantum Theory, 1918–1927,'" *Centaurus* 27 (1984), 76–99.

[7]Max Jammer, *The Conceptual Development of Quantum Mechanics* (New York: McGraw-Hill, 1966); Gerald Holton, *Thematic Origins of Scientific Thought: Kepler to Einstein* (Cambridge, Mass.: Harvard University Press, 1973, 1988).

[8]Specifically: Philipp Frank, *Einstein: Sein Leben und seine Zeit* (Munich: Paul List, 1949), published in English as *Einstein: His Life and Times*, tr. George Rosen, ed. Shuichi Kusaka (New York: Alfred A. Knopf, 1947); Anton Reiser (pseud. of Rudolf Kayser), *Albert Einstein: A Biographical Portrait* (New York: Albert and Charles Boni, 1930); and Carl Seelig, *Albert Einstein: Eine Dokumentarische Biographie* (Zurich: Europa Verlag, 1954). One must include as well as Einstein's own fascinating intellectual autobiography, in Schilpp, *Albert Einstein: Philosopher-Scientist.*

[9]*New York Times Sunday Magazine,* 15 December 1996.

[10]Tetsu Hiroshige, "The Ether Problem, the Mechanistic World View, and the Origin of the Theory of Relativity," *Historical Studies in the Physical Sciences* 7 (1976): 3–82.

[11]Gerald Holton, *Einstein, History, and Other Passions,* 174–175.

[12]Gerald Holton, *Introduction to Concepts and Theories in Physical Science* (Cambridge, Mass.: Addison–Wesley, 1952), 506; Robert A. Millikan, "Albert Einstein on His Seventieth Birthday," *Reviews of Modern Physics* 21 (1949), 343–344.

[13]Holton, *Thematic Origins of Scientific Thought,* ch. 8 and 477–480.

[14]Cited in Robert S. Shankland, "Conversations with Albert Einstein," *American Journal of Physics* 31 (1963):47–57.

[15]Seelig, *Albert Einstein: Eine Dokumentarische Biographie,* 61–62.

[16]Banesh Hoffmann, with Helen Dukas, *Albert Einstein: Creator and Rebel* (New York: Viking, 1972).

[17]Lewis S. Feuer, *Einstein and the Generations of Science* (New York: Basic Books, 1974).

[18]*New York Times,* 16 November 1919, 8.

[19]At the 1832 meeting of the Association of German Scientists and Physicians in Vienna, for instance, a hymn proclaimed, "Laßt uns vereint der Isis Tempel bauen / Der Göttin, welcher keine andre gleich / Die rätselhaft so nahe uns und ferne / Im Sandkorn thront wie dort im Flammensterne." See H. Schipperges, *Weltbild und Wissenschaft: Eröffnungsreden zu den Naturforscher-versammlungen 1822 bis 1972* (Hildesheim: H. A. Gerstenberg, 1976).

[20]Einstein to Besso, 13 May 1917; from Albert Einstein and Michele Besso, *Correspondance, 1903–1955,* tr. and intro. Pierre Speziali (Paris: Hermann, 1972), 114.

[21]Albert Einstein, Hedwig Born, and Max Born, *Briefwechsel, 1916–1955* (Munich: Nymphenburger Verlagshandlung, 1969), 215.

[22]Robert K. Merton, *Science, Technology, and Society in Seventeenth Century England,* 238.

[23]For a classic exposition of the contrast between *Kultur* and *Zivilisation,* see N. Elias, *Über den Prozess der Zivilisation,* 2 vols. (Basel: Verlag Haus zum Falken, 1939), vol. 1, 1–42.

[24]See, for example, *Brockhaus Enzyklopädie,* 1990.

[25]Bruno Gebhard, *Handbuch der Deutschen Geschichte,* 8th ed. (Stuttgart: Union Verlag, 1962), vol. 3, 305.

[26]Karl Mannheim, *Ideology and Utopia: An Introduction to the Sociology of Knowledge* (translation of *Ideologie und Utopie,* 1929; tr. Lewis Wirth and Edward Shils) (New York: Harcourt, Brace & World, 1970), 156. By 1843 Karl Marx had noted (in his *Kritik des Hegelschen Staatsrechts*) that "Geld und Bildung" were the main criteria for social differentiation in the *bürgerliche* society; Marx, *Karl Marx, Friedrich Engels, Werke,* ed. Institut für Marxismus-Leninismus beim ZK der SED (Berlin: Dietz Verlag, 1957), vol. 1, 203-333. For a useful summary of the *Bildungsbürgertum,* see F. Gregory, "Kant, Schelling, and the Administration of Science in the Romantic Era," *Osiris* (second series) 5 (1989):17–35.

[27]In the original sense of the educated Mandarinate that served the Chinese Empire and were chiefly concerned with administering or furthering the political and social needs of the state authorities. Fritz K. Ringer, *The Decline of the German Mandarins: The German Academic Community, 1890–1933* (Cambridge, Mass.: Harvard University Press, 1969).

[28]Christa Kirsten and Hans-Jürgen Treder, eds., *Albert Einstein in Berlin, 1913–1933,* 2 vols. (Berlin: Akademie-Verlag, 1979), vol. 1, 207. This document collection contains reports to the German foreign office from German diplomats in The Hague, Oslo, Copenhagen, Paris, Buenos Aires, Tokyo, Madrid, Montevideo, Rio de Janeiro, Chicago, New York, and Vienna. See vol. 1, 225–240.

[29]K. Mannheim, *Ideologie und Utopie* (Frankfurt: Verlag G. Schulte-Bulmke, 1969: first published in 1929), 221–222.

[30]For Albert Einstein's family tree, see Aron Tänzer, "Der Stammbaum Prof. Albert Einsteins," *Jüdische Familien-Forschung: Mitteilungen der Gesellschaft für jüdische Familienforschung* 7 (1931): 419–421.

[31]Erik Erikson, "Psychoanalytic Reflections on Einstein's Centenary," in Gerald Holton and Yehuda Elkana, eds., *Albert Einstein: Historical and Cultural Perspectives* (Princeton, N.J.: Princeton University Press, 1982), 151–173.

[32]Maja Winteler-Einstein, "Albert Einstein—Beitrag für sein Lebensbild," *Collected Papers: The Collected Papers of Albert Einstein* (Princeton University Press, 1987), vol. 1, xlviii–lxvi.

[33]Einstein, "Autobiographical Notes," 3–5.

[34]Nevertheless, Helen Dukas insisted that Einstein's lifestyle in Zurich and Bern was "anything but 'bohemian,'" as noted in L. Pyenson, *The Young Einstein: The Advent of Relativity* (Bristol: Adam Hilger, 1985), 77, note 9. Pyenson would have characterized Einstein not as a rebel but a stranger or marginal man (60–61).

[35]Seelig, *Albert Einstein: Eine Dokumentarische Biographie*, 125. The graduate student's name was Hans Tanner; Einstein supervised Tanner's dissertation while a professor at Zurich University.

[36]Giuseppe Castagnetti and Hubert Goenner, "Directing a Kaiser-Wilhelm-Institut: Albert Einstein, Organizer of Science?" paper given at the Boston University Colloquium for Philosophy of Science, 3 March 1997.

[37]Cited in Otto Nathan and Heinz Norden, eds., *Einstein on Peace* (New York: Schocken, 1968; reprint of 1960 edition), 157.

[38]Letter to Hans Muehsam, 30 March 1954, Einstein Archive 38–434; cited in *The Quotable Einstein*, ed. Alice Calaprice (Princeton, N.J.: Princeton University Press, 1996), 158. See also Einstein's declaration of his religiosity in Harry Graf Kessler, *Tagebücher 1918–1937*, ed. Wolfgang Pfeiffer-Belli (Frankfurt: Insel-Verlag, 1961), 521–522, and in Hubert Goenner and Giuseppe Castagnetti, "Albert Einstein as Pacifist and Democrat During World War I," *Science in Context* 9 (1996): 348–349. See also Albert Einstein, *Ideas and Opinions* (New York: Dell, 1954).

[39]Ministerium der geistlichen, Unterrichts- und Medizinalangelegenheiten, "Lehrpläne und Lehraufgaben für die höheren Schulen, nebst Erläuterungen und Ausführungsbestimmungen" (Berlin: Wilhelm Hertz, 1892), 20.

[40]Reiser, *Albert Einstein: A Biographical Portrait*, 26. Toward the end of his life, when Einstein's sister Maja visited him in Princeton (as Einstein wrote to Besso), both would spend their time together reading "Herodotus, Aristotle, Russell's *History of Philosophy*, and many other interesting books." Einstein and Besso, *Correspondance*; see also Albrecht Fölsing, *Albert Einstein: Eine Biographie* (Frankfurt: Suhrkamp, 1993), 819. One might add here that Heine was often excluded from the "official" cultural canon, especially outside Jewish circles, because of his religious background and his affiliation with French and revolutionary ideas.

[41]See the introduction by Maurice Solovine to Albert Einstein, *Letters to Solovine* (New York: Philosophical Library, 1987), 8–9. Auguste Comte is

notably absent from Einstein's reading list or exchanges. Comte remained relatively unknown in the German-speaking parts of Europe at the turn of the century. German translations of his works were slow to appear; see the chronology in Auguste Comte, *Rede über den Geist des Positivismus*, tr. and intro. I. Fetscher (Hamburg: Felix Meiner Verlag, 1994; originally published in 1844), xliii–xliv. In 1914, none other than Wilhelm Ostwald translated Comte's *Prospectus des travaux scientifiques nécessaires pour réorganiser la société*, almost a century after it was first published in 1822.

[42] Max Talmey, *The Relativity Theory Simplified, and the Formative Period of its Inventor* (New York: Falcon Press, 1932), 164.

[43] Seelig, *Albert Einstein: Eine Dokumentarische Biographie*, 17. The course was Professor Stadler's lecture course on "Die Philosophie I. Kants"; see *Collected Papers: The Collected Papers of Albert Einstein* (multiple vols.; Princeton, N.J.: Princeton University Press, 1987–), vol. 1, 364.

[44] Albert Einstein, "Elsbachs Buch: Kant und Einstein," *Deutsche Literaturzeitung* 1 (n.f.), 1685–1692.

[45] Immanuel Kant, *Critique of Pure Reason*, tr. Norman Kemp Smith (London: Macmillan, 1929), 113.

[46] Albert Einstein, "Remarks Concerning the Essays Brought Together in this Co-Operative Volume," in *Albert Einstein: Philosopher–Scientist*, 674.

[47] This data base for all books remaining after his death was compiled by NHK (Japan Broadcasting Corporation) and is scheduled to be published. A reading list of additional books may be found in Abraham Pais, *'Subtle is the Lord . . .': The Science and the Life of Albert Einstein* (Oxford: Oxford University Press, 1982).

[48] Max Weber, *Wissenschaft als Beruf* (Berlin: Duncker & Humboldt, 1967), 37; cf. Isaiah Berlin, *The Crooked Timber of Humanity*, ed. Henry Hardy (New York: Vintage Books, 1992), 213–216.

[49] As Fritz Stern shrewdly observed in a passage mentioning both Goethe and Einstein: "A genius could also be seen as a public nuisance. . . ." Stern, *Dreams and Delusions: The Drama of German History* (New York: Alfred A. Knopf, 1987). See also the first chapter of this work, on "Einstein's Germany." On the uses and abuses of Goethe by German ideologues, as well as on how Einstein's view of himself as a Jew differed from others (e.g., Fritz Haber), see Stern, *The Politics of Cultural Despair: A Study in the Rise of the Germanic Ideology* (Berkeley: University of California Press, 1961).

[50] Seelig, *Albert Einstein: Eine Dokumentarische Biographie*, 88–89.

[51] The letter was printed in Holton, *The Advancement of Science, and its Burdens: The Jefferson Lecture and Other Essays* (Cambridge: Cambridge University Press, 1986), 65; it is also hinted at in a letter to Max Born, cited in Born, "Physics and Relativity," *Helvetica Physica Acta, Supplementum IV* (1956), 249.

[52] *Collected Papers*, vol. 2, 150.

[53] Issac Newton, *Mathematical Principles of Natural Philosophy* (translation of *Philosophiae naturalis principia mathematica*), 2 vols., original translation by

Andrew Motte (1729), revised translation by Florian Cajori (Berkeley, Calif.: University of California Press, 1962), vol. 2, 398.

[54]Holton, *The Advancement of Science, and its Burdens*, 15.

[55]The theme of unity and unification also played an important role in biology, as Vassiliki Smocovitis has documented in her *Unifying Biology: The Evolutionary Synthesis and Evolutionary Biology* (Princeton, N.J.: Princeton University Press, 1996). William Morton Wheeler commented (as cited in Smocovitis, 109) that it might take "a few super-Einsteins" to unify biology, using Einstein as the icon of the theme of unification.

[56]Cited in Holton, *The Advancement of Science, and its Burdens*, 86.

[57]Abraham Pais, *"Subtle is the Lord . . ."*, 9.

[58]Pauline Mazumdar, *Species and Specificity: An Interpretation of the History of Immunology* (Cambridge: Cambridge University Press, 1995).

[59]Gerald Holton, *Science and Anti-Science* (Cambridge, Mass.: Harvard University Press, 1993), 12–15.

[60]Jürgen Renn and Robert Schulmann, "Introduction," in *Albert Einstein— Mileva Marić: The Love Letters*, ed. Renn and Schulmann (Princeton, N.J.: Princeton University Press, 1992), xi–xxviii.

[61]See Matthew Arnold, *Culture and Anarchy*, ed. Samuel Lipman (New Haven, Conn.: Yale University Press, 1994; first published in 1869).

[62]Einstein kept sculptured busts of both Goethe and Schiller in his Berlin home. F. Herneck, *Einstein privat: Herta W. erinnert sich an die Jahre 1927 bis 1933* (Berlin: Buchverlag Der Morgen, 1978), 47–48.

[63]In the latter part of the nineteenth and in the early twentieth century, it was quite common to assemble "best book" lists of the outstanding works of literature. In 1911, Heinrich Falkenberg compiled such a bibliography, "Listen der besten Bücher," in the *Zeitschrift für Bücherfreunde*; it comprised forty-six entries. The earliest such bibliography was Johann Neukirch's *Dichterkanon* of 1853; in Neukirch's compilation, as well as in the subsequent ones, Goethe played a dominant role. Around 1906, the Viennese bookseller Hugo Heller polled a number of intellectuals about their choice of the "ten best books." A selection of the responses was printed in the *Jahrbuch deutscher Bibliophilen und Literaturfreunde*, ed. H. Feigl (Zurich: Amalthea-Verlag, 1931), 108–127. As one might expect, Goethe figured prominently in these replies, both explicitly and implicitly. At that time, the consensus about the classic literary canon was so strong that it almost went without saying. Much has changed since then. Some ninety years later, the German weekly *Die Zeit* again asked a group of German intellectuals about the literary canon. This time they were to nominate only three to five works that they thought German *Gymnasiasten* had to read. Goethe, and particularly his *Faust*, still received numerous nominations, but now many respondents lamented the almost complete erosion of the classic literary canon. Indeed, this project of *Die Zeit* was intended to help resurrect a canon that had clearly faded. The report started with the statement that nowadays "up to 90 percent of those who begin to study German at a university do not know *Faust*"—

which to the earlier generations of *Bildungsbürger* would have sounded utterly unbelievable. *Die Zeit*, 16 May 1997.

[64] *Collected Papers,* vol. 1, 26–27.

[65] George Henry Lewis, *The Life of Goethe*, 3rd ed. (London: Smith, Elder and Co., 1875).

[66] Henry C. Hatfield, *Goethe: A Critical Introduction* (New York: New Directions, 1963), 28.

[67] And of course not only in German-speaking countries; to cite a single example, Ralph Waldo Emerson taught himself German specifically in order to read Goethe's works. See Robert D. Richardson, *Emerson: The Mind on Fire* (Berkeley, Calif.: University of California Press, 1995).

[68] Johann Wolfgang von Goethe, *Goethes Werke* (Hamburger Ausgabe), 4th ed. (Hamburg: Christian Wegner Verlag, 1962), vol. 13, 7–10.

[69] Hatfield, *Goethe: A Critical Introduction,* 114.

[70] Isaiah Berlin, *Concepts and Categories* (New York: Viking Press, 1979).

[71] Lewes, *The Life of Goethe,* 558.

[72] On the other hand, there can be no doubt, of course, that many of the *Bildungsbürger* and of those aspiring to their ranks rampantly quoted from this and all other classics merely to demonstrate their membership in the educated elite. Such people were greatly helped by Georg Büchmann's *Geflügelte Worte: Der Zitatenschatz des deutschen Volkes,* 27th ed. (Berlin: Haude & Spenersche Buchhandlung, 1926), a best-selling compilation of classic quotations and lengthier excerpts that was first published in 1864, and went through 27 editions by 1926. See Wolfgang Frühwald, "Büchmann und die Folgen: Zur sozialen Funktion des Bildungszitates in der deutschen Literatur," in *Bildungsbürgertum im 19. Jahrhundert,* part II: *Bildungsgüter und Bildungswissen,* ed. Reinhart Koselleck (Stuttgart: Klett-Cotta, 1990), 197–219.

[73] Walter Moore, *Schrödinger: Life and Thought* (Cambridge: Cambridge University Press, 1989), 47.

[74] Arnold Sommerfeld, *Electrodynamics* (volume three of his *Lectures on Theoretical Physics,* tr. Edward Ramberg) (New York: Academic Press, 1952), 311.

[75] In a September 1900 letter to Marić. *Collected Papers,* vol. 1, 260.

[76] Taken from Stewart Atkins's prose translation, *Johann Wolfgang von Goethe: Faust I & II* (Cambridge, Mass.: Suhrkamp/Insel Publishers Boston, 1984), lines 430–431, 434–441, 447–448.

[77] Maxwell's equations in empty space are taken from E. M. Purcell, *Electricity and Magnetism,* 2d. ed. (New York: McGraw–Hill, 1985), 331. I am fairly sure no physics text would connect them today with Dr. Faust, who is under a dark cloud these days for his various transgressions. See Roger Shattuck, *Forbidden Knowledge: From Prometheus to Pornography* (New York: St. Martin's Press, 1996).

[78]In Boltzmann's epigraphs, his rendering of Goethe's lines was, for Part I of his treatise: "So soll ich denn mit saurem Schweiss/Euch lehren, was ich selbst nicht weiss." For Part II, Boltzmann wrote: "War es ein Gott, der diese Zeichen schrieb, / Die mit geheimnissvoll verborg'nem Trieb / Die Kräfte der Natur um mich enthüllen / Und mir das Herz mit stiller Freude füllen."

[79]There is no authoritative picture of how Goethe imagined that heavenly Sign of the Macrocosmos, since no stage directions for it appear in the text of *Faust*. There is, of course, a good amount of literature on that question; see, e.g., Ernst Beutler, ed., *Johann Wolfgang von Goethe: Die Faustdichtungen* (Munich: Winkler Verlag, 1977), 754–757; Heinrich O. Proskauer, ed., *Goethes Faust: Erster Teil* (Basel: Zbinden Verlag), 1982; Rudolf Steiner, *Geisteswissenschaftliche Erläuterungen zu Goethes Faust* (Freiburg: Novalis-Verlag, 1955), vol. 1, 25–27 and *Die Rätsel in Goethes "Faust": exoterisch und esoterisch* (Dornach, Switzerland: Rudolf Steiner Verlag, 1981); and Erich Trunz's two editions, *Goethes Faust* (Hamburg: Christian Wegner Verlag, 1949), 496–497, and *Goethe—Faust* (Munich: C. H. Beck, 1986), 517–518). But we know at least the image that seems to have satisfied Goethe himself: in the 1790 edition of volume 7 of his writings (which included *Faust*), he commissioned as a frontispiece a version of an etching by Rembrandt, which historically was known as representing "Dr. Faust," named after the original sixteenth-century legendary figure (as shown in figure 4).

[80]Max Planck, "Die Einheit des physikalischen Weltbildes," in his *Vorträge und Erinnerungen* (Darmstadt: Wissenschaftliche Buchgesellschaft, 1970), 28–51.

[81]Originally entitled "Motiv des Forschens," the address was published under the rather unfortunate title "Prinzipien der Forschung" in Einstein's *Mein Weltbild* (Frankfurt: Ullstein Bücher 1955, first published in 1934), 107–110. This led to "Principles of Research" in the English translation of the address in *Ideas and Opinions* (New York: Dell, 1954), 219–222. While the second quote from this address is here taken directly from the published English translation, the first is my own translation of the original German text (in *Mein Weltbild*).

[82]David Cassidy, *Einstein and Our World* (Atlantic Highlands, N.J.: Humanities Press, 1995), 14.

[83]Anne Harrington, *Reenchanted Science: Holism in German Culture from Wilhelm II to Hitler* (Princeton, N.J.: Princeton University Press, 1996), 5, 10.

[84]See, for example, Peter Galison and David J. Stump, eds., *The Disunity of Science: Boundaries, Contexts, and Power* (Stanford, Calif.: Stanford University Press, 1996) and Ian Hacking, "Disunified Sciences," in Richard Q. Elvee, ed., *The End of Science? Attack and Defense* (Nobel Conference XXV; St. Peter, Minn.: Gustavus Adolphus College, 1992), 33–52.

[85]Henry Adams, *The Education of Henry Adams: An Autobiography* (Boston: Houghton Mifflin, 1918).

Peter Galison

The Americanization of Unity

I T WAS LATE 1946. RETURNING FROM MOBILIZATION, scientists around Cambridge—as elsewhere in the United States—were streaming back to the university. Philipp Frank, who had helped usher in the scientific philosophy of the Vienna Circle and was now a lecturer in the Harvard department of physics, set out a plan for Warren Weaver at the Rockefeller Foundation entitled "The Institute for the Unity of Science: Its Background and Purpose." It is immensely tempting and indeed historically useful to read this manuscript backwards, to see in it the tree whose seed had been planted in late-night discussions at the *Arkadenkaffee,* to track its manifold roots back to the early 1920s in Berlin and Vienna. On such a reading, the revised, now American, Unity of Science movement would chiefly be a revivification of the older Viennese one. Surely there were common concerns: both movements sought to rid philosophy of "superfluous" metaphysics and replace it with a clarity, precision, and empiricism for which science provided the template. Indeed, both in the prewar and postwar Unity of Science efforts, modern science, and not only physics, loomed large. There was near-unanimity that Boltzmann, Mach, Einstein, and Bohr had done much more than rewrite the rules for physics; they had set a new agenda for philosophy. Observability, causality, and probability now reigned where *Geist und Volk* once had.

This essay, however, will take up the new Institute for the Unity of Science that emerged in postwar America, not exclusively through its distant root-ends, but in its immediate envi-

Peter Galison is the Mallinckrodt Professor of History of Science and of Physics at Harvard University.

ronment. It will not focus so much on the scientist-philosophers of the interwar German-speaking world of modernism and Marburg neo-Kantianism, but rather on the squadrons of American scientists returning in 1945 from war research that had given science both a new form of work and a novel place for physics, chemistry, engineering, psychology, and sociology in the world. The objective, in short, is to elicit a double vision: a picture of postwar unity that is both the extension of the Vienna Circle and, at the same time, a philosophical outlook squarely located in the scientific concerns of an age of computers and nuclear power.[1]

Gerald Holton, a participant from early on, has recalled that the assembly of the American Unity of Science movement at Harvard began with Philipp Frank's organization of an "Inter-Scientific Discussion Group" in 1944. The group rapidly expanded, with such speakers as the polymathematician Norbert Wiener, biophysicist John Edsall, and sociologist Talcott Parsons coming to talk about a wide range of topics from biophysics and computers to the psychoanalysis of social systems.[2] Even before the war was over, Frank and his colleagues began dreaming of a new Institute for the Unity of Science.

Beginning with the familiar lament that science had grown ever more specialized, Frank gestured in his December 1946 report toward those who argued that every attempt at integration would descend into superficiality. True, there were those, including Harvard President James Conant, who were concerned about the political consequences of an education insufficiently wide to undergird liberal democracy. But Frank was worried: with quack prescriptions for unity lurking on one side, and popular science and Hollywood movies calling from the other, good scientists understandably wondered where to turn. Working against this fragmentation, Frank contended, was another, deeper tendency. In the world of the late twentieth century, "cross connections" were growing, not shrinking: "The domains of facts which can be derived from one and the same set of principles have not become smaller but larger." As the cross-connected scientists pulled fields together, domains of the special sciences "merged." Chemistry and physics provided an example. Fifty years earlier, no physicist could truly understand chemistry; now, physical chemistry and chemical physics had entered the scene: "Today general chemistry is just a part of nuclear physics. The physicist has an easy road into the very heart of chemistry." So it was in geometry, where general relativity guided the physicist into the heart of mathematics. And just as mathematical biophysics had joined biology and physics, behaviorism had sealed the union between psychology and biology. Throw in F. S. C. Northrop's unification of political and religious ideologies and their links with the physical sciences, Frank contended, and one was on the way to a universal *pass-partout* at the physicists' disposal.[3]

As Frank represented the problem, one difficulty was that of language. While the various special fields of science held much in common, the bridges between them were blocked by gross and fine differences in meaning that were unfortunately confused with differences over matters of fact. "The situation reminds [us of] the Biblical story of the tower of Babel. Because of the confusion in human language the tower of science cannot grow into . . . heaven." Logical positivism, now given a less Viennese and more cosmopolitan pedigree, was advanced in America by J. B. Stallo and Charles Sanders Peirce, William James, and John Dewey, and on the Continent by such luminaries as Henri Poincaré and Ernst Mach; now the whole (according to Frank) had been cast into a more "modern" formulation in work on both sides of the Atlantic. Percy Bridgman introduced operationalism, and Charles Morris bound the Americans to the core of the Vienna Circle that included Otto Neurath, Rudolf Carnap, Ludwig Wittgenstein, and Moritz Schlick.[4] In the classical (1930s) formulation of the unification project, Frank and his allies had been after a semantical goal, above all: to show that the special sciences could all be put into a language of everyday life. This continued in some versions of Frank's postwar philosophy.[5] Now, in this 1946 program, Frank wanted more—a "socio-psychological analysis" or "pragmatic" approach to supplement the logico-empirical one that previously had been the exclusive goal. As Frank put it:

> By adapting these approaches a vast field of research is opened up. "Hybrid fields" like "mathematical biophysics" or "math-

ematical economics" are no longer isolated cells where some awkward professors may enjoy their strange fancies but by the application of logico-empirical and socio-psychological analysis these "cross-connections" become the roots of new developments leading towards the integration of human knowledge and human behavior. These queer cross-connections become the avanguards [sic] of the science of the future.[6]

Only by such bonds could the investigation establish the connection between "contemporary physics on one side and contemporary religion and politics on the other side with contemporary philosophy being the intermediate link."[7] In another document from the same time, Frank listed some of the goals of a sociology of science—it would include the conditions under which discoveries were made, but also "intervention of the government in science," and "contemporary merging of science and technique."[8]

Warren Weaver had thrown the weight of the Rockefeller Foundation (along with some modest resources) behind the prewar Unity of Science movement run by Neurath, Carnap, and Morris. In the much-changed postwar world, Weaver heard Frank out and recorded in his diary on December 13, 1946 that "the Unity of Science Movement has been in a somewhat chaotic state since the death of Otto Neurath [late in the war], this being the more true since N[eurath] ran all of the business of the organization in a very individualistic and indeed almost dictatorial way."[9] To the old commitments of the Unity of Science movement (an encyclopedia, a journal, a bibliography, and conferences on unified science) Frank now wanted to add the role it might play in "modern American movements in general education." Writing to Weaver in January of 1947, Frank explained that his course aimed to show just what the "principle of relativity" sanctioned in the wider world of culture, ethics and truth—and what it did not. Only through such a critical examination could the scientist know whether the theory of quanta justified the belief in "freedom of the will" or advanced the reconciliation of science and religion. So girded against misinterpretation, the student of science could venture out against the raft of pseudoscientific or the pseudoreligious interpretations of science. When coupled with an understanding of the historical situatedness of science, such as the Copernican Revolution and "similar conflicts," Frank contended that the science student would have an "inner track" in grasping current relations between science, religion, and government.[10]

At stake, Frank argued, was the fate of the world. Ideologies—combinations of philosophical with political creeds—underpinned both the right wing with its organismic metaphysics and the left wing with its dialectical materialism. Prominent "cardinals of the church" espoused their Thomism (so Frank continued), while political leaders including Lenin plunged his followers into dialectical materialism. Only the student with logico-empirical analysis in one hand and socio-psychological analysis in the other could navigate these waters, for only with a deep understanding of the scientific process in context could the student grasp the idea that a chemical formula like H_2SO_4 was not an isolated fragment of knowledge but rather a "flaming manifesto to mankind."[11]

Weaver bought the manifesto. Recorded among the foundation's deliberations are the considerations that moved them. Above all, the Board cited the ever-expanding "cross- and inter-connections" between pairs of disciplines that now seized "more and more common ground": physics and chemistry, astronomy and physics, biology and psychology, among others. These, the panel judged, were "domains of experience . . . explainable from one and the same set of basic principles." Accordingly, in December 1947, the Rockefeller Foundation designated some $9,000 for the Unity of Science movement covering three years of support (though it was not, for technical reasons, delivered until July of 1949).[12] Led by directors Rudolf Carnap, Charles Morris, Philipp Frank, Milton Konvitz (a lawyer from Cornell), and Hans Reichenbach (then at the University of California, Los Angeles), the group took every opportunity to proclaim their limitless ambition—they would re-establish ties with Europe, train a generation of politically astute scientists, link the working scientific disciplines together, and reform philosophy.

Could this unification take place? If so, would it reflect a unified nature or a unified science? A confidential Rockefeller Foundation report to the trustees (dated March 1949) meditated on this metaphysical dilemma. "We have physical experiments, chemical experiments, biological experiments, and other specialized techniques, but it is important to remember that classification into these categories is man's invention. Whether it is also nature's, we don't know." One school of scientists, the report continued, supposed such metaphysical unity did obtain: "a universe of matter and energy whose interactions under certain conditions produce motion, radiation, and the other effects which we label physical, and under different conditions produce the nightingale's singing and other behavior which we call biological."[13] From Alfred North Whitehead to George Sarton, this metaphysical commitment to the unity of nature became an oft-repeated creed.[14]

Not everyone agreed, as the foundation's 1949 report made clear. Herbert Dingle, for one, argued that this sort of reductionistic metaphysical unity could not be guaranteed. The Rockefeller trustees would have read in the report that the metaphysical unity of nature was not a sure thing, according to Dingle:

> We aim at it; we hope we shall achieve it; but we must recognize the possibility that nature may be essentially dual, or even multiple. . . . We do not ignore the organic unity of nature when we consider laws of motion apart from those of economics, let us say. We simply avail ourselves of the fact that we can make progress by admitting that, at present, motion and economics are disconnected subjects of study. We hope that we shall unify them, but to let our thinking be influenced by the assumption that they are essentially one seems indefensible.[15]

That said, the report went on to laud Maxwell's unification of electricity and optics, along with Einstein's of mathematics and physics (through general relativity). But the list did not stop there. Of crucial import were biophysics, biochemistry, psycho-physics, psycho-physiology, and social psychology; moreover, the report noted, "other borderland sciences are fields that seem likely to contribute new data for a unitary picture of nature." In the process of this joining together of "borderland" disciplines in pairwise links, concepts that were superfluous would drop by the wayside. Einstein's geometrical dynamics made "gravitational force" a dead letter; the quantum theory of the chemical bond rendered "chemical force" obsolete; and Maxwellian electrodynamics left fundamental optical hypotheses as nothing but a fifth wheel.[16] Would this *piecewise* integration extend all the way from mathematics to sociology? If it did, would the knowledge pyramid reflect a "natural" order of things? Steering a midcourse between metaphysical dualism and metaphysical unity, Herbert Feigl argued for establishing such connections "without premature attempts at complete unification."[17]

Partial connections (such as that afforded by chemical physics) would take place through the "master key" of semantics, "the study of the meaning of words and other symbols." Just as disposing of "chemical force" was a conceptual advance, so too would be a clarification of the myriad of often obsolete terms plaguing biology—"entelechy," "vital force," "mechanism," "holism," and "entity"—not to speak of similar vestiges of an earlier physics, including "absolute space," "absolute time," "simultaneity." Only a rigorous operationalism could effect this purge of the superfluous. Quoting Feigl approvingly, the report continues, "The possibility of a reconstruction of all factual sciences on the basis of a common set of root terms enables us to speak of the reducibility of all sciences to a common, unitary, interscientific language."[18]

In an attempt to deliver just such a "basic operational dictionary," Frank and MIT's Karl W. Deutsch began a composition in the fall of 1952.[19] Containing nineteen different categories, with three hundred terms, the sweep of the project is stunning.

Table 1. Frank and Deutsch, Basic Operational Dictionary (Outline)

I. Basic Notions	XI. Physiological Concepts
II. Sets, Groups, Order, Structure	XII. Organism
III. Constructs (of physics)	XIII. Mechanism
IV. Prediction	XIV. Learning
V. Logic and Semantics	XV. Biology
VI. Psychology	XVI. Ethics
VII. Communications Engineering and Theory	XVII. Religion
VIII. Sociology and Anthropology	XVIII. Chemistry
IX. Economics	XIX. Aesthetics
X. Political Science	

Source: Frank, "Report on the Dictionary of Operational Definitions" (September 1952, RG 1.1, 100 Unity of Science, 1952–56, Box 35, folder 285, Rockefeller Foundation Archives).

In all, there would be three hundred "basic concepts." These would include not only standard physics notions like mass, matter, energy, space, time, and field but also (picture Carnap's horror) such hard-to-imagine-operationalized concepts as love (under psychology) or God, belief (faith), soul, and damnation (under religion). (I cannot help but wonder here whether salvation is excluded deliberately or whether it is operationalized under a negative disposition of damnation.)

In those cases where the operational definitions were clear from usage, they would be drawn from "scientific writing." If not, then views would be drawn from writers with an appropriate "operational viewpoint." If both were absent, then experts would provide "paper and pencil operations"; if even these were not possible, then "hypothetical operations," analogous to procedures that could be performed, would be utilized. By example: "real" might refer to that which is "familiar from repetitive, gross, bodily experience." Alternatively (Frank wrote), "we mean by 'real' things from which we can continue to learn, overriding past symbols and traditions." "Reality" is signaled by "structural coincidence" between sensations and impersonal records. "Sensations" track back to "traces" within the nervous system and are therefore impermanent and not easily verifiable, whereas "instrument records" are external, more easily verifiable, and forever.[20]

One could study these three hundred greatest hits in the concept parade almost mechanically, finding here and there the bits and pieces of prewar Vienna Circle concerns. Starting with "sets, groups, order, and structure," one could discern the elements of the new formal logic and set theory of Frege and Russell that so impressed the group back in the 1920s: class, universals, group, model, order, congruence make their appearance here. Under "prediction" we could track back many of Reichenbach's or Carnap's concerns in their extensive writings on probability: "equipossibility," "limit of relative frequencies," "degree of assent," or Frank's own youthful dissection of the causality notion that had so impressed Einstein. Here, too, we find vestiges of the old Vienna Circle's fascination with Freudian psychology (the list includes id and ego) and the frequently discussed *gestalt* concept that arose in discussions among Carnap, Neurath, Wittgenstein, and Schlick; we also see elements of economic theory (utility, market, profit, labor, capital, efficiency) that engaged many among the left wing of the Circle. Religious concerns, anathema to Carnap, could no doubt be laid (in part) at Charles Morris's door, as his "Paths of Life" drew him ever more into contemplation of the great world religious leaders and their thought.[21]

Then, too, we are not surprised to find on a list drawn up by Frank, the positivistic biographer of Einstein, the terms "mass," "matter," "energy," "space," or "time," under "constructs (of physics)." These Einstein-revised notions were read by the Vienna Circle as prototypical positivistic moves. Space was defined through the laying out of rigid measuring rods, and time by the readings of identically calibrated clocks; notions of mass and energy were correspondingly revised. As we now know, Einstein demurred when presented with Frank's positivistic rendition of his work; as far as I can tell these protestations were to no avail.[22]

But there is more on Frank's list than its vaulting ambition, more even than the sum of prewar interests. In particular, several of the categories are *not* ones we would have found even among Neurath's wildest hopes. "Communications engineering and theory" had no role in the world of Schlick, Carnap, Neurath, or Frank years before. This category breaks down as shown in Table 2.

Table 2. Communications Engineering and Theory

1. Message
2. Information
3. Signal
4. Channel
5. Circuit
6. Network
7. Recognition
8. Noise

Source: Frank, "Communications Engineering and Theory" (1952, RG 1.1, 100 Unity of Science, 1952–56, Box 35, folder 285, Rockefeller Foundation Archives).

While pieces of this list were discussed together in various sectors of the radio-technical or telephonic industries before the war, their pride of place in the basic concepts of the world is altogether new. And for just cause—these are some of the starting points of the new sciences loosely grouped under cybernetics, informatics, and the burgeoning wartime radar laboratories. By the time Frank composed this list in September of 1952, cybernetics had already become one of the central issues preoccupying the meetings of the Institute for the Unity of Science. On January 19, 1951, Wiener and Rosenblith launched the Institute's "Cybernetics and Communications" study group with two meetings devoted to a "systematic examination of fundamental concepts," beginning with feedback, noise, entropy, and information.[23] Just to give a few examples of the scope of the study group's concerns, consider the sessions organized around the linguist Morris Halle on the "Entropy of Language," building on the formal ties Claude Shannon and Wiener had established between entropy in statistical mechanics as $k \log \Omega$ and information defined as negative entropy. (Wiener himself had published directly on the links between cybernetics and these other fields in his "Speech, Language, and Learning."[24]) Two further meetings of the Institute also fit squarely into the Wienerean framework: R. D. Luce spoke on "Communication and Learning in Small Task-Oriented Groups," and M. Rogers addressed "Some Applications of Information Theory to Psychology." Indeed, the cybernetic track remained one of the most active topics at the Institute for several years. Many of these gatherings took place in the department of electrical engineering at MIT.[25] Why, the reader may wonder, would anyone be discussing small-group learning, language, and psychology in MIT's E.E. department? In a sense, the answer to that question lies at the bottom of a fundamental change in the meaning of "unity" in the Institute's conception of "Unity of Science." But to get there we need to step back, both philosophically and historically.

* * *

The central point here is that the Central European prewar notions of unity differ strikingly from the ideas of unity that emerged from American collaborative war work. Before the war the slogan "unity of science" carried various meanings, as we know from the careful work of Nancy Cartwright, Jordi Cat, Richard Creath, Michael Friedman, Lola Fleck, Thomas Uebel, and others.[26] Carnap, for example, had at least two senses of the notion of unity. In his early work there is a kind of autopsychological foundationalism, the search for an *Aufbau* built upon the "bedrock" certainty of one's sense impressions here and now, and encompassing pyramidically the whole of human knowledge, including psychology. In a sometimes uneasy tension with this notion of a whole structure for science lies another in which the "basis" is not the autopsychological but rather the heteropsychological, the commonly shared experience. Upon this, too, Carnap could erect the whole edifice of the sciences. Finally, as Friedman has so interestingly shown, Carnap seems—at least by the time he wrote *Logical Syntax of Language* (1934) and perhaps the *Aufbau* itself—to have a full-blown conventionalism that allows him to make it a matter of indifference how the *Aufbau* is grounded: the whole point is to secure objectivity by way of the relations between its elements, and to avoid any reliance upon private experience.[27]

Neurath's notion of unity is not the same as any of these. It is not a "building up" (*Aufbau*) on rock-bottom autopsychological foundations, it is not an *Aufbau* on heteropsychological foundations, and it is not a conventional or "structural" *Aufbau* either. As Cartwright, Cat, Fleck, and Uebel note, Neurath's most distinct notion of unity aimed not at the expression of all science in the language of physics, but at the creation of a heterogeneous jargon capturing pieces of social science, ordinary language, and physics.[28] Metaphorically, he did not want a pyramid of knowledge but a coordinated encyclopedia. The hierarchy of a Comtian picture of science would be replaced by the orchestration of different instruments, each distinct but brought together to accomplish something bigger than any could do individually.[29] Famously, Neurath invoked the image of a forest fire to illustrate how necessary it was to organize the various sciences into an effective unit. Surely, Neurath insisted, one would need to know about climatological and chemical laws to understand the pattern of a mass conflagration. But without the coordination of these physical laws with sociological ones, there could be no prediction. Presumably, the "orchestrated" effort would survey human behavior to find out the circumstances in which humans tended to cause fires. Only then, chemical and climatological knowledge in hand, would the coordinated science be able to predict how the fire spread.[30]

The sciences of World War II were worried about massive fires, all right, but they were in the business of causing them—not reckoning their probability of accidental occurrence. More importantly, American scientific war work was characteristically not merely aggregative (chemical knowledge + climatological information + sociological aspects of behavior = knowledge of fire propagation) but instead involved the formulation of entirely new combinations of disciplines. Take Norbert Wiener, who participated in many of the Institute's meetings and whose work (as we have seen) launched the Institute's central and sustained inquiry into the new science of cybernetics. Long before World War II, Wiener had been a precocious and nearly omnicompetent scholar who moved easily between his home field of mathematics and the adjacent ones of Birkhoff's ergodic theorem, Kolmogorov's axiomatization of probability theory, and a variety of philosophical inquiries. The German air assault on Britain changed that. More than a year before Pearl Harbor, Wiener threw himself completely into the problem of antiaircraft fire–control; suddenly he was neck-deep in engineering and vacuum tube work. Looking for a collaborator in March of 1941, he saw no need for a pure mathematician—he required someone already immersed in computing, communication engineering, and vacuum tube work. Anyone without a feel for engineering, without at least competence in putting together radio sets, ought not apply: "There is nothing in abstract algebra or topology . . . which would prepare one in any way to cooperate in engineering design."[31] And this (according to Wiener) was not only true for his project; it was equally so for just about every piece of crucial war work, from ballistics to cryptography. Already—almost a year before America's war had begun—Wiener insisted that science had to remake itself, realigning old subjects and creating new ones.

Swerving erratically through the sky as they skirted flak, Nazi bombers were hard to hit; a shot from the ground took ten or fifteen seconds to reach altitude, and by then the bomber was headed somewhere else. Responding to this difficulty, Wiener launched what became a new science, one devoted to the electromechanical replication of the human capacity to predict: "Since our understanding of the mechanical elements of gun pointing appeared to us to be far ahead of our psychological understanding, we chose to try to find a mechanical analogue of the gun pointer and the airplane pilot."[32] With electronics, Wiener radically reorganized the way gun laying was done by putting together a "predictor," a device that would take the measure of the pilot's last moves, calculate the statistical likelihood of his future actions, and launch a shell to destroy him where he would be. In a sense that rapidly became altogether explicit for Wiener, the antiaircraft predictor precisely duplicated the intention of the pilot in the flow of electrons. Psychology was not being *superadded* to electromechanism; psychological notions were *supplanted* by the circuitry. Immediately, Wiener began considering what this picture of psychology would mean.

Edwin Boring, the Harvard psychologist (and an early member of the Inter-Scientific Discussion Group) took note. On November 13, 1944, he wrote to Wiener that he planned a "pretty complete list of psychological functions" that he hoped Wiener could duplicate by means of electrical systems. "Symbolic process" would, in electrical input/output terms, come down to "a delayed, adequately differential reaction"; "introspection" on the electronic breadboard ultimately became a reaction to a reaction. Boring laid down the challenge: could Wiener translate each of these stimulus-response pairs into his own "black-box" of electrical relations? "Generalization"? "Abstraction"? Each term that Boring put into behaviorist form would, he hoped, then find its expression in circuitry. "I do not know that you can [do it], but I should be betting on you."[33]

With lightning speed Wiener generalized his predictor's supplanting of intention. Even before the war was over, he had begun to make fundamental and broad use of terms previously confined to specialized aspects of telephonic engineering: signal, noise, feedback, and control flew far ahead of antiaircraft fire.

The physiological model was rife with feedback systems—a circumstance brought home to him by his medical collaborators. There was, for example, the clinically well-known purpose tremor in which a voluntary act such as reaching for a pencil launches an uncontrollable oscillation of overshoot and undershoot. As reformulated by Wiener, the purpose tremor became a particularly salient instance of the more general functioning of the brain—in this case a disordering of the normal feedback cycle between brain, muscle, effector organ, outside world, receptor organ and back to the brain.[34] This work linked him crucially to a variety of medical personnel, including Walter Cannon at the Harvard Medical School, an active member of the Institute for the Unity of Science. At the same time, Wiener became increasingly interested in the role such feedback circuitry might play in electronic computing. With John von Neumann, Wiener organized early and important meetings that helped usher in plans for computers being formulated toward the end of the war.

These discussions were pivotal for von Neumann. In 1945 and 1946, von Neumann did his fundamental work on the digital computer, abstracting from the particular form of electronic realization of the system and putting together a brainlike composite of "organs," leading to the stored-program computer. Wiener, von Neumann, and their associates moved back and forth between the language of logic, the language of electronics, the language of neurophysiology, and the abstracted language of computer functions. Much of this—including von Neumann and Goldstine's first lecture on flow diagrams—came together in January of 1947 at the Harvard Symposium on Large-Scale Digital Calculating Machinery. H. H. Aiken, builder of the electromechanical Mark I and one of the leading experts on the computer, brought the results a few days later to the Inter-Scientific Discussion Group.[35] Here were the new "borderland" sciences in action.

This working out of a conjoint picture of behaviorist psychology and the feedback-predictive circuitry explored by Wiener, neurological studies, and electronic computation became an enduring concern for the Unity of Science group. The same day that Boring penned his fourteen-concept challenge to Wiener, Boring reported in another letter that he and Wiener had both just been present at the Inter-Scientific Discussion Group.[36] Wiener himself was an active member of the group, speaking at one of the very first gatherings in December of 1944 on Birkhoff's ergodic theorem and again on February 14, 1945, on "The Brain and Computing Machine": it was not accidental that the Institute had elevated cybernetics into one of its central topics.[37] The discipline of "cybernetics" (named after the war by Wiener) set out and emphasized certain concepts such as feedback and control, gave it a more developed formal presentation, and linked the whole to information theory and computational strategies. With enormous, perhaps overwrought enthusiasm, physiologists, sociologists, anthropologists, computer designers, and philosophers leapt on the cyberwagon. Even the anthropologists Margaret Mead and Gregory Bateson rewrote the framework of their work in light of the new concepts.[38] Recalling Frank and Deutsch's 1952 list of basic concepts in communication theory (message, information, signal, channel, circuit, network, recognition, and noise), we now perceive in it the elements of shared starting concepts. We see precisely the kind of piecewise unification and cross-connections proclaimed by the Institute for Unified Science.

At the close of the war, Frank had only to look around the corridors of Cruft Laboratory in his own physics department to find his colleagues, fresh from the war effort, brimming with enthusiasm about the new interdisciplines to be explored. Frank himself had spent part of the war preparing Navy officers in physics for their work with radar—along with E. C. Kemble, I. Bernard Cohen, Gerald Holton, Roy Glauber, and Frederic de Hofmann. In the latter part of the war, Frank moved to Columbia University where he did classified applied mathematics work. Among Frank's other physics department colleagues, each had his own stories, his own witnessing of disciplinary recombination. Indeed, their interdisciplinary duties were typically several. The acoustician F. V. Hunt ran the multimillion-dollar Harvard Underwater Sound Laboratory, drawing together electronics, oceanography, physics, ship operation, and much else.[39] Wendell Furry had codirected a research project on the thermal diffusion of gases subject to molecular force laws, work executed on the differential analyzer at MIT and eventually put to application on the Manhattan project.[40] E. C. Kemble had been on the Alsos mission to determine how far the Nazis were in their quest for the bomb; his conclusion was that the ability of American scientists to retrain themselves into engineers had been the key to a narrow victory over the Germans. It was the Allies' great fortune, Kemble concluded, that the Nazis had remained rigidly hierarchical and protective of their division of the pure and applied disciplines.[41] (Both Kemble and Alsos leader Samuel Goudsmit attended the early Inter-Scientific meetings.) Kenneth Bainbridge was returning from the Manhattan Project, out of which the new field of "nucleonics" was to combine nuclear physics, engineering applications, and a myriad of electronic, chemical, metallurgical, mathematical, calculational, and even medical techniques. Now many of these physicists joined forces with colleagues from such fields as metallurgy, chemistry, and engineering to run Harvard's interdisciplinary Committee on Nuclear Sciences.

In the Manhattan Project itself, physicists had learned how to think about matter very differently. There was no way to avoid seeking simultaneously to understand the metallurgy, shock wave behavior, and nuclear physics of imploding plutonium. It is against this background that we must read Frank's 1946 plea cited at the beginning of this paper: "Today general chemistry is just part of nuclear physics. The physicist has an easy road into the very heart of chemistry." In 1928 the specialty "nuclear physics" did not exist as such; the utterance would have been meaningless.

Physicists Edward Purcell, Wendell Furry, Curry Street, and Julian Schwinger spent their war years based further down Massachusetts Avenue at the MIT Radiation Laboratory, where quantum mechanicians had made common cause with radio engineers and industrialists to produce the new field of microwave physics. Once physicists had held themselves aloof from the grubby details of engineering; now a new generation of American physicists learned from radio engineers how to think about black boxes, input/output analysis, effective circuits, breadboards, signal-to-noise ratios, and mass production. Fueled first by the war, from Stanford to MIT microwave physics burgeoned where before the war no such field had existed. Whether one looked up to radioastronomy or down to particle accelerators, whether one turned to the practical features of solid-state physics or to the moments probed by nuclear magnetic resonance, the new techniques of war-inspired short-wavelength physics were reformulating how people went about their scientific business and with whom they spoke. So when Purcell came to his first Inter-Scientific Discussion Group in March of 1946, interscientific coordination would have been more than a programmatic gesture; it had been his main work for the bulk of a still-young scientific career.[42]

And the list goes on: Harvard physicist John Van Vleck (also a member of the Institute for the Unity of Science) spent the years of conflict working on radar countermeasures to foil German air defense in Harvard's Radio Research Laboratory, far from his usual specialty and in concert with engineers. I. I. Rabi, a leader of the overall radar effort, also became a member of the Unity of Science movement. The high-pressure experimentalist Percy Bridgman, inventor of operationalism and one of the strongest American prewar boosters of the Unity of Science movement (having also served on the Inter-Scientific Discussion Group steering committee), spent his war years at the ORDC Ordnance section, where he worked with the Watertown Arsenal on the pressure effects of projectiles on steel, on polymers of new plastics for possible use inside internal combustion engines, and on the physics and chemistry of explosions and incendiary explosives.[43] Nor was war-driven interdisciplinary coordination restricted to physicists, chemists, and biologists. For his part, philosopher W. V. Quine, another powerful supporter of the prewar Unity of Science movement, joined other philosophers, engineers, and mathematicians to decipher intercepted messages and pinpoint the location of the German submarine wolfpacks.[44] And astronomer Harlow Shapley, who ran a substantial research effort on complex new lenses for aerial photography, went on to become a member of the board of trustees for the Institute for the Unity of Science. Indeed, one only has to look at the roster of the Inter-Scientific Discussion Group to find Cambridge addresses that had not existed before, addresses that joined fields as well as people: the Electro-Acoustic Laboratory, the Systems Research Laboratory, the MIT Radiation Laboratory, the Psycho-Acoustic laboratory, the Underwater Sound

Laboratory, the Fatigue Laboratory. Looking back at the list of participants in the Inter-Scientific Discussion Group and the Institute for the Unity of Science in this context, suddenly it reads differently.

Take Stanley S. Stevens, the Harvard psychologist, active in the Inter-Scientific Discussion Group and then on the board of trustees of the Institute for Unified Science and on its program committee.[45] Stevens had worked in and with a variety of inter-disciplinary wartime laboratories; it is worth pausing to review what they were and how they worked. Back in late 1940, when the War Department was first gathering project titles for imme-diate research, it turned certain problems over to the National Defense Research Committee. One issue raised by officers at Wright Field was the effect of noise and vibration on aviators. Information had come back that the noise and vibrations of airplanes so exhausted the pilots on their deep-penetration bombing missions over Germany that they crashed on landing back in East Anglia. Leo Beranek, an acoustician, got the job of develop-ing measures to address the airplane noise problem, first by deciding how quiet airplanes needed to be and then by develop-ing or finding lightweight materials to do the job. It is perhaps illustrative of the times that when Beranek proposed a budget of $3,000 per year, the request was denied—as Beranek recalled in April of 1945, the military indicated that if "we would multiply the figure by ten, [they] would talk business."[46] Forty thousand dollars were set aside for the first seven months. In addition to Stevens (from psychology) and Beranek, F. V. Hunt represented the Harvard physics department and the acoustician P. M. Morse came from MIT. It was a complex problem, involving the analy-sis of the principles of sound absorption and then its develop-ment into a workable product—fiberglass AA, produced by Owens-Corning Fiberglass company. Next the group began a systematic analysis of how to predict noise levels from the blueprints of planes, so that engineers could intervene early enough to mini-mize the problem. By April of 1941, the sound team had also begun to explore communication in airplanes, a task at once electrical (analysis of amplifiers and microphones), acoustical (insulation of headsets) and psychological (determination of which sounds were intelligible, development of codified patterns of speech). Two laboratories, both interdisciplinary, collaborated extensively in this effort: the Electro-Acoustical Laboratory (directed by Beranek) and the Psycho-Acoustic Laboratory (headed by Stanley Stevens).[47] These labs did not always find the coordination of different fields easy; as one wartime report put it, "The bringing together of men of different experience, different training and different interests presented a sizable problem of integration."[48] That "integration" was hard won over the next year and a half. But by 1944, interscientific collaboration had become part of everyday life for Stevens and his associates.

Problems addressed by the psycho- and electro-acoustic labs included designing earphones, headphones, and microphones in oxygen masks that would allow radio communication while not impeding visibility or oxygen flow; positioning a microphone for best communication in a noisy environment (not, it turned out, near the throat as the Army Air Forces were doing—even the hand-held microphone worked better). While the majority of the electro-acousticians scrambled to ensure quiet in the cockpit and on the bridge, Gerald Holton, secretary of the Inter-Scientific Discussion Group, spent his war days teaching radar while mixing acoustics, chemistry, and physiology in designing gas masks that wouldn't silence a sergeant's barked orders.[49] Physiology as well as psychology entered the picture when the electro-acoustic and psycho-acoustic labs began to address the problem of communication at high altitude; as the bombing levels rose in 1942, it became apparent that inter-crew communication was failing above 25,000 feet. Electronics experts, radio wizards, and doctors joined together to solve the problem. Volunteer conscientious objectors were (virtually) lofted in depressurization chambers to examine the effects of rarified atmosphere on the human voice. The results of these experiments showed that the problem lay in neither the equipment nor the faculty of hearing; instead, it was learned that the human voice drops to a mere 1/60th of its sea-level intensity when the speaker reaches 35,000 feet. Combined with new instruments and a new, codified set of mandatory speech protocols, interphone talk soon became considerably clearer. By the time Stevens began attending the Inter-Scientific Discussion Group on December 18, 1945, he would have recognized interdisciplinary (systems) coordination as both essential and effective.

Building on L. J. Henderson's and Elton Mayo's prewar fatigue laboratory, the war years saw an extension of studies to a wider domain of scientific fields and a broader spectrum of military applications. In 1943, an interdisciplinary team constructed a "fatigue chamber" at the Harvard Business School and began to make operational the myriad of physio-psychological effects associated with noise, all the while working to alter airplane and communication-gear design. How, precisely and repeatably, did the noise in a cockpit or on the bridge of a battleship impede response time, the execution of multiple tasks, or attentiveness to spoken orders? Addressing questions of this last sort fell to another laboratory, the Systems Research Laboratory, tasked with simulating a ship's combat information center in the heat of battle. Again, the task was irreducibly interdisciplinary as time and motion experts had to learn to work with radio engineers, physicists, psychologists, and radar-display personnel. Out of the chaos of a ship under attack, the team created a new order: they moved instruments, altered display panels, choreographed physical movements, and rewrote patterns of speech. This was a "systems" laboratory, where "systems" meant for participants a focus not on isolated pieces of equipment but on people operating equipment "as an integral part of an organization." These researchers were not aiming to discover how a particular radar component handled frequency, pulse width, repetition rate, and lobe pattern; they were after answers to other questions: When operated by typical personnel, how many target fixes can the radar handle per minute? What is the normal degree of error in these fixes? How much time lag is there between the appearance of the pip on the scope and the dissemination of range and bearing by the operator? What are the details—such as the location and size of controls, and the types of cursors—that delay getting each fix?[50]

Conjoint questions like these forced a specific form of unity among the various disciplines, a unity predicated on assembling diverse methods, professions, and patterns of work into the production of pragmatic solutions to immediate problems. Again and again, these interscientific laboratories rendered "operational" their solutions to applied wartime problems, at every step comingling psychological categories, physical principles, and engineering practicalities.

* * *

Put before your mind's eye two very different pictures. First, evoke the Vienna Circle in full bloom. Its adherents saw themselves as opposing the forces of irrationality and joining hands with the modern in architecture, city planning, and, at times, Austro-Marxist politics.[51] Never politically powerful or even institutionally secure, as time went on their voices were increasingly drowned out by the array of nationalistic forces pitted against them. The drive to a "Unity of Science Movement" was, for Neurath, Carnap, and their allies, part and parcel of a struggle to bring together a rationality and objectivity that would halt racial and nationalistic assaults from dominating the world. Their opponents were Austrian clericalism, entrenched traditional philosophy, and, later, Nazism. Just the title of a typical philosophy paper in the mid-1930s shows just how much metaphysics the "old" philosophy could cram into a single article: "Godliness and the Character of the '*Volk.*'" Whether through an *Aufbau* of Carnap's sort or through a physicalist thing language, the Vienna Circle's goal was to squeeze out of the world of the meaningful all that counted as metaphysics. And metaphysics was for them not some limited concept, but the alive, well, and dangerous movements for Godliness, *Volk*, mysticism, and *Deutschtum.* Even the philosophy of Heidegger, they believed, was infected by metaphysics.[52]

On the side of rationality was, above all, the new science and logic. And among the sciences, none served better and more epigrammatically than Einstein's 1905 paper on the electrodynamics of moving bodies. Indeed, if I had to choose one moment in the history of science that the Vienna Circle would have emblazoned on their banner, it would no doubt have been that most famous of all lines penned by that twenty-six-year-old patent clerk: "The introduction of a 'light-aether' will prove to be superfluous. . . ." That unapologetic stripping down, not unlike the Bauhaus architects' removal of ornamentation in their

Dessau headquarters, was for the Vienna Circle a move towards victory over everything they detested in philosophy, in politics, and in culture. Modern physics could ground itself in the specifiable, measurable world of function—and so, the left wing of the Vienna Circle believed, could the rest of the social and human sciences. Theirs was, as they often insisted, more than a philosophical movement; it was the search for a new *Lebensgestaltung*.

The unity in the prewar "unity of science" movement had to be papered over to a certain extent. Carnap's conception of his unifying scheme changed over time. Protocol statements, often deployed as essential in securing unity, were understood differently by Carnap and Neurath, with Neurath always insisting that he only meant that these statements were the last to be given up. Even when unity was to be a purely linguistic exercise— unity as expressibility in terms of objects describable in space and time—there were differences of understanding. Sometimes this language appears as the language of physics, sometimes (especially in Neurath) it refers to a mixed jargon, drawing, without possibility of reduction, from many different sciences (including the social sciences). But variously as the programs for scientific unity were construed, there was a shared sense that the project of *Erkenntniswissenschaft* would find a new and better formulation. Philosophy would aid the other disciplines in cutting the unnecessary or destructive and identifying the modern strategy of epistemic austerity.

Now move your mental image ahead to 1947, from Vienna and Berlin to Cambridge, Massachusetts. The scientific banner flying overhead is not that of relativity and quantum mechanics—though these might occasionally be invoked by Frank. Instead, the banner announces the riveting new, war-boosted interdisciplines: cybernetics, computation, neutronics, operations research, psycho-acoustics, game theory, biophysics, electro-acoustics. The old enemies of interwar Vienna are gone or vanquished: Austrian clericalism and the hollow vestiges of the Habsburg empire do not figure very large in Cambridge, and fascism has been slain, in no small measure (in their scientists' eyes, at least) because of scientists' intervention. Now these same tools that had won the war promised the world. Cybernetics, with its nonlinear feedback, was celebrated as offering a way to rewrite the social sciences as well as the sciences; the computer's logic was thought to be universal and capable of doing everything from weather forecasting to nuclear-weapons design, from the resolution of longstanding problems in number theory to modeling the human mind.

The unification these scientists had in mind was a unification through localized sets of common concepts, not through a global metaphysical reductionism. Were the mathematical and technical features of feedback, control, black boxes, flow diagrams, or extensive forms of a game "reducible" to nuclear physics? Hardly. Even posing that question about the kinds of problems facing the Institute seems hopelessly inappropriate. With the kind of power these scientists felt they had at war's end, fretting about ontological reductionism must have seemed almost beside the point. As the chemist E. Bright Wilson wrote to Holton, the Institute secretary, in 1950: "The phase of the Institute's work in which I am particularly interested is that which deals with scientific method in its most practical and least philosophical senses."[53] The Americanization of unity just after World War II was not sited around an isotypic picture language, a physical language, an *Aufbau*, or an orchestration. It was planted around the new sciences of Los Alamos, the MIT Radiation Laboratory, the stored-program computer of the Institute for Advanced Study in Princeton; this was to be a science unified in pieces, grounded in common widely applicable concepts, and promising a power beyond dreams.

One last contrast: When the Vienna Circle faced off against theology in their manifesto, they saw mystic obscurantism as a rising threat; however misunderstood or powerless they were, the Vienna Circle aimed to cast millenia of such speculation to the winds. When the organizers of the Institute for the Unity of Science sent out its first flyers, they made "Science and Faith" and "Science and Values" early and longstanding objects of study.

In one of the first meetings of the Institute for the Unity of Science, a prominent participant probably spoke for many in observing that the public now saw scientists as authorities comparable to the high priests of ancient cults. But the truly staggering feature is not the prominent positive role accorded truth and values; it is that in these first months of the *pax Americana*, this group of scientists, humanists, and philosophers could take on God and Morality as problems—and fully expect to solve them.

ACKNOWLEDGMENTS AND SOURCES

It is a pleasure to acknowledge the assistance of Rockefeller Archive Center, especially Thomas Rosenbaum, the archivist. Documents from the Rockefeller Foundation Archives are identified in the notes by the designation "RFA". Gerald Holton provided records from his time as secretary of the Inter-Scientific Discussion Group (later the Institute for the Unity of Science). These are grouped into numbered folders and are referred to in the notes as "GHP". Documents identified as "LBP" are from the privately held papers of Leo L. Beranek. Finally I would like to thank the Harvard University Archives (HUA) for use of the papers of Sterling Dow. A shorter, earlier version of this essay appeared in the *Deutsche Zeitschrift für Philosophie*.

ENDNOTES

[1] From Carnap's handbook; cited in Holton, "From the Vienna Circle to Harvard Square: The Americanization of a European World Conception," in F. Stadler, ed., *Scientific Philosophy: Origins and Developments* (Boston: Kluwer Academic Publishers, 1993), 47–73.

[2] Holton, "On the Vienna Circle in Exile," in W. DePauli-Schimanovich, E. Köhler and F. Stadler, eds., *The Foundational Debate: Complexity and Constructivity in Mathematics and Physics* (Boston: Kluwer Academic Publishers, 1995), 269–292; Holton, "From the Vienna Circle to Harvard Square." Also, see P. Masani to Holton, 5 April 1991, in which Masani recalls getting Frank, Shapley, LeCorbeiller, Birkhoff, Leontief, Bridgman and Uhlenbeck together for what became the "Inter-Scientific Discussion Group" in "1943 or 1944."

[3] Frank, "The Institute for the Unity of Science: Its Background and Its Purpose." 1946, RG 1.1, 100 Unity of Science, Box 35, Folder 281, RFA.

[4] Wittgenstein himself, as is well known, was often anything but happy with the idea of being grouped together with the Vienna Circle.

[5] Frank, "Background and Purpose."

[6] Frank, "Background and Purpose," 11.

[7] Ibid.

[8] [Frank], untitled, 1945–46, folder 4, GHP.

[9] "Interview: WW [Warren Weaver], Friday, December 13, 1946, Professor Philipp G. Frank," RG 1.1, 100 Unity of Science, Box 35, Folder 281, RFA.

[10]Frank to Weaver, 7 January 1947, RG 1.1, 100 Unity of Science, Box 35, Folder, 281, RFA.

[11]Frank to Weaver, 7 January 1947, pp. 7, 42–43, RG 1.1, 100 Unity of Science, Box 35, Folder 281, RFA.

[12]Resolution RF 47131 set a start date of 1 January 1948; this was emended to 1 July 1949 in RF 49085, both in RG 1.1, 100 Unity of Science, Box 35, Folder 281, RFA. The delay was due to the difficulty in establishing the proper tax status for the organization.

[13]"Toward Integration of the Sciences," March 1949, excerpted from "Confidential Report to the Trustees," p. 6, RG 1.1, 100 Unity of Science, Box 35, Folder 281, RFA.

[14]Whitehead, Lowell Lecture 1925, published as *Science and the Modern World* (New York: Macmillan Co., 1937); see also George Sarton, *Introduction to the History of Science* (Baltimore, Md.: published for the Carnegie Institution of Washington by the William and Wilkins Co., 1947–1948). Both cited in "Toward Integration," ibid.

[15]"Toward Integration," 7.

[16]Ibid., 7–8.

[17]Ibid., 11.

[18]Ibid., 13.

[19]Frank to Weaver, 29 September 1952, RG 1.1, 100 Unity of Science, Box 35, Folder 285, RFA.

[20]Ibid.

[21]Morris, *Paths of Life* (New York: Harper and Brothers, 1942).

[22]On Einstein's protest that his work was not properly understood as purely positivistic, see Holton, "Einstein, Mach, and the Search for Reality," in *Thematic Origins of Scientific Thought: Kepler to Einstein*, rev. ed. (Cambridge, Mass.: Harvard University Press, 1988), 261–262.

[23]Steering Committee to participants, 11 January 1951, folder 3, GHP.

[24]Wiener, "Speech, Language, and Learning," in *Journal of the Acoustical Society of America* 22 (1950): 696–697, reprinted in Wiener, *Collected Works*, ed. P. Masani, 4 vols. (Cambridge, Mass.: MIT Press, 1976–), vol. 4, 200–201; original references to Shannon include "A Mathematical Theory of Communication," *Bell Systems Technical Journal* (3) and (427) (1948).

[25]Institute for the Unity of Science, AAAS, 20 January 1953, from "The Steering Committee, to The Cybernetics and Communications Group of the Institute for the Unity of Science," RG 1.1, 100 Unity of Science, Box 35, Folder 285, RFA.

[26]The secondary literature on the history of the Vienna Circle in general and on matters relating to "unity" in particular is vast and growing. For an introduction to the material see Galison and Stump, *The Disunity of Science* (Stanford: Stanford University Press, 1996); Thomas Uebel, ed., *Rediscovering the Forgotten Vienna Circle* (Dordrecht: Kluwer, 1991); F. Stadler, ed., *Scientific*

Philosophy (1993); R. N. Giere and A. Richardson, eds., *The Origins of Logical Empiricism*, Minnesota Studies in the Philosophy of Science, volume 16 (Minneapolis, Minn.: University of Minnesota Press, 1996); and N. Cartwright, J. Cat, L. Fleck, and T. R. Uebel, *Otto Neurath: Philosophy Between Science and Politics* (Cambridge: Cambridge University Press, 1996).

[27]M. Friedman, "Carnap's *Aufbau* Reconsidered," *Noûs* 21 (December 1987): 521–546; and R. Creath, "The Unity of Science: Carnap, Neurath and Beyond," in Galison and Stump, eds., *The Disunity of Science*.

[28]J. Cat, N. Cartwright, and H. Chang, "Otto Neurath: Politics and the Unity of Science," in Galison and Stump, eds., *The Disunity of Science*.

[29]N. Cartwright, et al., *Otto Neurath*, 167–188.

[30]Ibid, 173–74.

[31]Galison, "Ontology of the Enemy," *Critical Inquiry* (1994): 228–266, 234.

[32]Ibid., 240.

[33]Ibid., 247.

[34]Wiener, *Collected Works*, vol. 4, 169–73.

[35]Program Committee of Interscience Discussion Group to participants, 7 January 1947, folder 2, GHP; Herman Goldstine, *The Computer from Pascal to von Neumann* (Princeton: Princeton University Press, 1947), 267.

[36]Boring to Masani, 13 November 1944, folder 1, GHP.

[37]Masani to Wiener, 6 December 1944; Masani to Ducasse, 6 February 1945, both folder 1, GHP.

[38]Steve J. Heims, *The Cybernetic Group* (Cambridge, Mass.: MIT Press, 1991).

[39]See "History of Harvard's War Contracts," in Sterling Dow Papers, HUA, and Dow interview with Eric Arthur Walker, S.D., Associate Director of the Underwater Sound Laboratory, 18 June 1945, Sterling Dow Papers, HUA.

[40]"Statement of War Activities," 1 February 1946, W. Furry folder, Sterling Dow Papers, HUA.

[41]"When asked how the Germans did so well despite their failure to use academic scientists to advantage, Kemble observed that the analytical engineers in Germany came close to beating us, in spite of the fact that they didn't use their pure scientists to advantage. If they had had somewhat more intelligent overall leadership, things might have come out differently. These engineers had originally good academic training." Interview with Edwin Crawford Kemble, 22 June 1945, Kemble File, Sterling Dow Papers, HUA.

[42]Attendance sheets for the early meetings of the Inter-Scientific Discussion Group can be found in GHP, folder 2.

[43]Bridgman, "Two major jobs for OSRD, Others for Watertown Arsenal," 1 May 1945, P. Bridgman folder, Papers of Sterling Dow, HUA.

[44]On Quine's war work, see David Kahn, *Seizing the Enigma* (Boston: Houghton Mifflin Company, 1991), 241 ff.

[45]See, e.g., Program Committee to "Dear Sir" (participants), 24 October 1950, folder 3, GHP.

[46]Beranek, as narrated to S. Dow, "History of Electro-Acoustic Laboratory (Formerly called: Research on Sound Control)," 9, LBP.

[47]Beranek, "Electro-Acoustic Laboratory," ibid.

[48]Beranek, "Our Laboratory During the War," Report Covering Activities Between December 1, 1940, and July 1, 1944, Cruft Laboratory, Harvard University, Cambridge, Massachusetts, 9.

[49]See L. Beranek folder, Papers of Sterling Dow, HUA; Holton, informational communication, 28 March 1996.

[50]"Systems Research Laboratory: Wartime Organization and Purpose of the Systems Research Laboratory, Harvard University," 5, LBP. On the prewar Fatigue Laboratory, see Bard Clifford Cosman, "The Human Factor: The Harvard Fatigue Laboratory and the Transformation of Taylorism," Harvard Senior Thesis, 1983; Richard Gillespie, *Manufacturing Knowledge: A History of the Hawthorne Experiments* (Cambridge: Cambridge University Press, 1991).

[51]Galison, "Aufbau/Bauhaus: Logical Positivism and Architectural Modernism," *Critical Inquiry* 16 (1990): 709–752; Galison, "Constructing Modernism: The Cultural Location of Aufbau," in Giere and Richardson, eds., *Origins of Logical Empiricism*, 17–44.

[52]For a discussion of the fierce struggle between Carnap and Heidegger, see M. Friedman, "Overcoming Metaphysics: Carnap and Heidegger," in ibid., 45–79.

[53]E. Bright Wilson, Jr., to Holton, 6 October 1950, folder 3, GHP.

... [T]he result is derived that physics actually speaks about the mental phenomena of individuals. Then it is quite natural to interpret modern science in favor of an idealistic or skeptical world view and to deny that science can provide knowledge about our physical reality. In a similar way, in quantum theory, the impossibility of introducing position and velocity of a particle at a certain instant of time as state variables has also been interpreted by statements in common-sense language—by a "short circuit," without the long chain that leads to observable phenomena. In one such interpretation the position and velocity of a particle at the same instant of time are said to be inaccessible to the research abilities of human beings, and according to another interpretation, these quantities are actually not strictly determined but vague. If we use the common-sense meaning of the terms employed, this can only mean that the world itself is something vague and can be investigated, not by the methods of science, which are striving for precise and logical results, but by methods used in investigating the "irrational" and "spontaneous" aspects of the world: by metaphysics, religion, mysticism. So some have seen in modern physics a possibility of reconciling science with religion and metaphysics.

—Philipp Frank

"Contemporary Science and the
Contemporary World View,"
from *Dædalus* Winter 1958,
"Science and the Modern World View"

Lorraine Daston

Fear and Loathing of the Imagination in Science

ECENTLY A READER RESPONDED with dismay to a *New Yorker* article by historian Daniel J. Kevles about the charge of scientific fraud brought by Margot O'Toole against Thereza Imanishi-Kari. What distressed this reader was not so much the issue of fraud itself as Kevles's argument that the exercise of judgment and imagination in science was essential and should not be conflated with fraud:

> ... I am troubled by Kevles's acceptance of a need for scientists to be imaginative in analyzing research results. What might the public's realization that this practice exists do to its confidence in the hard sciences? Will we next be expected to believe that accountants require imagination in their work?[1]

Such expressions of uneasiness about the role of the imagination in science are not new. When the physicist John Tyndall delivered a "Discourse on the Scientific Use of the Imagination" to the British Association for the Advancement of Science in 1870, he too drew shocked reactions from the press. The London *Times* was severe:

> The glory of a Natural Philosopher appears to depend less on the power of his imagination to explore minute recesses or immeasurable space than on the skill and patience with which, by observation and experiment, he assures us of the certainty of these invisible operations. . . . [Tyndall] confesses that Mr. Darwin "has drawn heavily upon time and adventurously upon matter." We ask ourselves whether we are listening to one experimental philosopher

Lorraine Daston is Director at Max Planck Institute for the History of Science, Berlin.

describing the achievement of another experimental philosopher. We had been under the impression that Natural Philosophers drew no bills.[2]

The echo of fiscal analogies reverberates over the space of more than a century: scientists should be as methodical (and as plodding) as accountants ("Natural Philosophers draw no bills"). To permit the imagination to infiltrate science is to tamper with the books, to betray a public trust.

My aim here is not to show that first-rate science requires imagination; others have already pleaded this point with vigor and eloquence.[3] Rather, I would like to explore how and why large portions of the educated public—and many working scientists—came to think otherwise, systematically opposing imagination to science. I shall argue that the critical period was the mid-nineteenth century, when new ideals and practices of scientific objectivity transformed the persona of the scientist and the sources of scientific authority. More specifically, I shall focus on the apparent paradox, also first framed in the early decades of the nineteenth century, that the more scientists insisted upon the obduracy and intransigence of facts, the more they feared the power of their own imaginations to subvert those facts. Why would scientists convinced of the power of ugly facts to murder beautiful theories, as Thomas Henry Huxley famously put it, nonetheless take heroic precautions to protect those burly facts from gossamer-spun imagination?

The key to this paradox lies buried within the histories of the scientific fact, on the one hand, and of the faculty of the imagination, on the other. In order to dramatize the novelty of the mid-nineteenth-century developments, I shall begin with a brief account of how eighteenth-century natural philosophers and natural historians understood the relationship between scientific facts and the scientific imagination. The pivot of my story is the polarization of the personae of artist and scientist, and the migration of imagination to the artistic pole. At roughly the same time that artists working in a romanticist vein emphasized creativity over mimesis, scientists troubled by the overthrow of one time-honored theory after another in quick succession sought more durable achievements. This early nineteenth-century confrontation of individualistic, brashly subjective art with collective, staunchly objective science was not simply the collision of some timeless faith in the imagination with an equally timeless faith in facts. Rather, it signaled a mutation in the meanings both of imagination and of facts that still shapes the moral economy of science.

FRAGILE FACTS, NECESSARY IMAGINATION

Experience we have always had with us, but facts as a way of parsing experience in natural history and natural philosophy are of seventeenth-century coinage. Aristotelian experience had been woven of smooth-textured universals about "what happens always or most of the time"; early modern facts were historical particulars about an observation or an experiment performed at a specific time and place by named persons.[4] What made the new-style facts granular was not only their specificity but also their alleged detachment from inference and conjecture. Ideally, at least, "matters of fact" were nuggets of pure experience, strictly segregated from any interpretation or hypothesis that might enlist them as evidence. Some seventeenth-century philosophers were as skeptical as their twentieth-century successors about the bare existence of what we now (redundantly) call theory-free facts. René Descartes, for example, trusted only those experiments performed under his own supervision, because those reported by others distorted the results to "conform to their principles."[5] Even the most vigorous promoters of "matters of fact" acknowledged that these nuggets of pure experience were hard won: Francis Bacon thought only the strict discipline of method could counteract the inborn tendency of the human understanding to infuse observation with theory.[6] The 1699 *Histoire* of the Paris Académie Royale des Sciences confessed that the "detached pieces" of experience the academicians offered in lieu of coherent theories or systems had been wrenched apart by a "kind of violence."[7] Chiseling out "matters of fact" from the matrix of interpretation and conjecture was hard work.

But it was the hard work of smelting and purifying, not that of building and constructing. One of the most striking features of the new-style scientific facts of the seventeenth century is how swiftly and radically they broke with the etymology that connected them to words like "factory" and other sites of making and doing. In Latin and the major European vernaculars the word "fact" and its cognates derives from the verb "to do" or "to make," and originally referred to a deed or action, especially one remarkable for either valor or malevolence: *facere/factum, faire/fait, fare/fatto, tun/Tatsache*.[8] English still bears traces of this earlier usage in words like "feat" and, especially, in legal phrases like "after the fact." When the word "fact" acquired something like its familiar sense in the early seventeenth century as "a particular truth known by actual observation or authentic testimony, as opposed to what is merely inferred, or to a conjecture or fiction," to quote from its entry in the *Oxford English Dictionary*, it snapped the philological bonds that had tied it to words like "factitious" and "manufacture." Conversely, by the mid-eighteenth century, once-neutral words like "fabricate" (originally, to form or construct anything requiring skill) or "fabulist" (teller of legends or fables) had acquired an evil odor of forgery and deception in addition to their root senses of construction. For most Enlightenment thinkers, facts *par excellence* were those given by nature, not made by human art. "Facts" and "artifacts" had become antonyms, in defiance of their common etymology.

In keeping with the opposition of natural facts to human artifacts, the errors that most terrified Enlightenment savants in theory and practice were the errors of construction, of a world not reflected in sensation but made up by the imagination. Sensory infirmities worried Enlightenment epistemologists of science relatively little, prejudices and misconceptions instilled by bad education rather more so, the distortions wrought by strong passions still more, and the unruly creations of the imagination most of all. These latter seemed so pervasive as to make the simplest factual narrative a triumph of vigilance, discipline, and civilization in the minds of some Enlightenment writers. Bernard de Fontenelle, Perpetual Secretary of the Académie Royale des Sciences in Paris, thought the inclination to embellish the facts of the matter in any retelling so irresistible that "one needs a particular kind of effort and attention in order to say only the exact truth." It took centuries before society advanced to the point of being able to "preserve in memory the facts just as they happened," before which time "the facts kept in [collective] memory were no more than visions and reveries."[9]

The chronic inability to hold fast to fact, to keep the inventive imagination in check, was a midpoint along a continuum to madness. Scientists were as much at risk as poets from the diseases of the imagination. In Samuel Johnson's allegorical novel *The History of Rasselas Prince of Abissinia*, the philosopher Imlac meets a learned astronomer "who has spent forty years in unwearied attention to the motions and appearances of the celestial bodies, and has drawn out his soul in endless calculations." Upon further acquaintance the astronomer proves as virtuous as he is learned, "sublime without haughtiness, courteous without formality, and communicative without ostentation." Surely the astronomer is the long-sought-after happy man, content in his science and virtue? Alas, no; the astronomer is stark raving mad. He discloses to Imlac his delusion that he alone can control the world's weather and that he therefore bears the crushing responsibility for the welfare of the world's population on his shoulders. Imlac reflects that no one is immune from the depredations of the imagination: "There is no man whose imagination does not sometimes predominate over his reason, . . . All power of fancy over reason is a degree of insanity; . . . By degrees the reign of fancy is confirmed; she grows first imperious, and in time despotic. Then fictions begin to operate as realities, false opinions fasten upon the mind, and life passes in dreams of rapture or anguish."[10]

It was not only novelists and philosophers who worried about "fictions [that] begin to operate as realities," about the fragility of facts in the face of overweening imagination. Practicing naturalists also fretted openly. In his monumental *Mémoires pour servir à l'histoire des insectes* (1734–42) the French naturalist and experimental physicist René Antoine Réaumur warned that "although facts were assuredly the solid and true foundations of all parts of physics," including natural history, not all reported facts in science could be trusted. It was not simply a matter of weeding out hearsay or dubious sources; even sincere, well-trained naturalists could adulterate observations with imaginings. Citing the example of Godaert's observation that some insects could spawn insects of a different species, Réaumur preached caution: "Too often the observer has the disposition to see objects quite otherwise than they [actually] are. The extravagant love of the marvelous, a too strong attachment to a system fascinates his eyes."[11] An errant imagination was also Georges Cuvier's diagnosis of how Jean-Baptiste Lamarck had gone astray in natural history: for all of his scientific gifts, Lamarck was one of those minds that "cannot prevent themselves from mixing [true discoveries, *découverts véritables*] with fantastic conceptions. . . .[T]hey laboriously construct vast edifices on imaginary bases, similar to the enchanted palaces in our old romances which disappear when the talisman upon which their existence depends is broken."[12]

Cuvier's opposition of "true discoveries" to "romances," of fact to fiction, was at least as old as Bacon and was echoed countless times before, during, and after the Enlightenment. Equally banal and enduring was the parallel opposition of the faculties of reason and imagination. What was striking about eighteenth-century views of the imagination in light of later developments is their firm insistence that the imagination, despite its perils, was as essential to philosophy and science—the pursuits of reason—as to the arts. Moreover, both art and science drew on the same kind of healthy imagination—and both were at risk from the same pathologies of the imagination. Both science and art were, in the view of most of their eighteenth-century practitioners, dedicated to revealing the truths of nature; imagination enlisted to this aim was a sound, sane one, that is to say, an imagination subject to rules. Even the most inventive genius should, Enlightenment critics insisted, bow to the authority of nature and its rules. John Dryden, for example, wondered whether Shakespeare might not have gone too far in creating the monstrous character of Caliban in *The Tempest*, "a person which was not in Nature," and Goya explained the famous epigram of his *Caprichos*—"The sleep of reason produces monsters"—as a call to the union of reason and imagination: "Imagination deserted by reason produces impossible monsters. United with reason, imagination is the mother of the arts and the source of their wonders."[13]

Images of the monstrous pervaded Enlightenment accounts of the diseased imagination in the arts and sciences. Voltaire distinguished between the "active imagination," which inspired the finest works of mechanics, mathematics, poetry, and the fine arts, and the "passive imagination," which caused violent passions, fanaticism, delusions, and monsters both figurative and literal. The passive imagination in the arts and sciences welded together "incompatible objects" into chimeras; in the womb of a pregnant woman it could impress the soft embryo with the form of some hideous perception—for example, of a convicted criminal broken on the wheel—received by the mother.[14] The French critic Jean-François Marmontel acknowledged that fiction was no servile imitation of nature, but even fiction that perfected nature still kept the imagination on a short leash. What was variously called the "marvelous," "monstrous," or "fantastic" imagination in the arts led to the "debauchery of genius."[15] Poets and artists were instead directed to obey the cardinal rule of verisimilitude: "A verisimilar fact is a fact possible in the circumstances where one lays the scene. Fictions without verisimilitude, and events prodigious to excess, disgust readers whose judgment is formed."[16] Enlightenment good taste demanded that even fictions be decked out as possible facts and that art as well as science follow nature. Both art and science required imagination, but in neither should the imagination be allowed to invent at will.

Or rather, *against* will, for in the view of Enlightenment writers like Voltaire and Marmontel the pathological imagination overthrew the reasonable sovereignty of the will. Whereas the healthy, active imagination always partakes of judgment and "raises all of its edifices with order," the diseased, passive imagination acts imperiously, so that its victims are no longer "master" of themselves.[17] Here the distinction between the healthy and the diseased imagination took on moral as well as epistemological (and aesthetic) undertones. The consequences of submitting weakly to the domineering imagination could be dramatic, as the members of the French scientific commission formed in 1784 to investigate alleged phenomena of animal magnetism emphasized. After observing the remarkable convulsions and cures displayed by mesmerized patients, the commission—which included the astronomer Jean-Sylvain Bailly, the chemist Antoine Lavoisier, and the electrician Benjamin Franklin—decided to undergo animal magnetism themselves. Seated in the great mesmeric tubs, under the magnetizer's wand, the commissioners contrasted their own calm impassivity with the spectacular crises of the *convulsionnaires*:

> The Commissioners could not help but be struck by the difference between the public treatment and their own particular treatment in the tubs. The calm and silence of the one, the motion and agitation of the other; there, the multiple effects of violent crises, the habitual state of mind and body interrupted and troubled, nature exalted; here, the body without pain, the mind untroubled, nature preserving both its equilibrium and its ordinary course, in a word the absence of all effects.[18]

Tranquil and self-controlled savants versus shaking and shrieking patients: for the commissioners there could be no clearer contrast between the sound and the diseased imaginations. They concluded that the cures wrought by animal magnetism were often genuine, and the convulsions mostly sincere, but that all were the work of the imagination, "that active and terrible power that produces the great effects that one observes with astonishment at the public treatment."[19] Although gender and class played some role in how the commissioners gauged degrees of susceptibility to the imagination, the ultimate defense against "that active and terrible power" was enlightenment (*lumières*), a combination of intelligence and self-mastery. Despite their palpable disapproval of such excesses of the imagination, the savants of the Royal Commission paid tribute to its extraordinary power over mind and body. No romantic poet was ever more firmly convinced of the force of the unfettered imagination than the Parisian savants.

HARD FACTS, WILD IMAGINATION

In Enlightenment art and science, the imagination was Janus-faced: on the one hand, it was essential to creative work in both realms; but on the other, it could betray the natural and the verisimilar by breeding monsters. Its power verged on the supernatural. It could drive brilliant artists and scientists mad, it could trigger violent seizures, it could cure the hopelessly ill, it could distort and obliterate facts. So long as art and science shared a common goal of truth to nature, they also shared a code of aesthetic, epistemological, and moral values that praised one face of the imagination and deplored the other. Genius—be it in poetry, sculpture, or natural philosophy—was the expression of heightened imagination. Whether the genius in question was Milton or Leibniz, Michelangelo or Descartes, the natural endowment that made their achievements possible was in essence the same: a soaring imagination that "produces more than it discovers . . . [that] hatches brilliant systems or discovers great truths."[20] Imagination was not yet immiscible with science, and it was arguably more robust than facts.

Between about 1780 and 1820 this configuration changed dramatically. Put in the briefest terms, facts hardened, the imagination ran riot, and art and science diverged in their aims and their collective personae. Within the narrow confines of this essay, it is only possible to offer emblematic episodes to illustrate the nature and extent of these major transformations in the self-images of artists and scientists. Immanuel Kant's account of genius in his *Kritik der Urteilskraft* heralded things to come. Kant took it for granted that originality was the *sine qua non* of genius and that "[e]veryone is agreed on the point of the complete opposition between genius and the *spirit of imitation*." But that which can be learned, reasoned Kant, can also be in a sense imitated. Hence even the greatest triumphs of the natural sciences could no longer count as true works of genius:

> So all that Newton has set forth in his immortal work on the *Principles of Natural Philosophy* may well be learned, however great a mind it took to find it all out, but we cannot learn to write in a true poetic vein, no matter how complete all the precepts of the poetic art may be, or however excellent its models. The reason is that all the steps that Newton had to take from the first elements of geometry to his greatest and most profound discoveries were such as he could make intuitively evident and plain to follow, not only for himself but for every one else.[21]

Kant was second to none in his admiration for Newton and the revelations of the natural sciences, but he nonetheless denied even Newton the title of genius. For Kant, the very transparency and communicability of mathematics and the natural sciences removed them from the realm of profound, ineffable originality inhabited by Homer or even Christoph Wieland. Kant's emphasis on communicability linked the natural sciences to an emergent opposition between objectivity and subjectivity that Kant himself pioneered. Kant employed these terms in several distinct senses in his critical philosophy; I wish to draw attention here only to the sense that resonated most loudly for nineteenth-century scientists and that meshed most tightly with Kant's rejection of the bare possibility of scientific genius. In the closing pages of the *Kritik der reinen Vernunft*, Kant offered a rough-and-ready test for distinguishing objectively valid convictions from merely subjectively valid persuasions:

> If the judgment is valid for everyone, provided only he is in possession of reason, its ground is objectively sufficient [*objektiv gültig*], and the holding of it to be true is entitled conviction. If it has its ground only in the special character of the subject, it is entitled persuasion. . . . The touchstone whereby we decide whether our holding a thing to be true is conviction or mere persuasion is therefore external, namely, the possibility of communicating it and of finding it to be valid for all human reason.[22]

In the middle decades of the nineteenth century this ideal of objectivity as communicability, shorn of every idiosyncrasy and particular perspective, was realized in the emergence of international, long-term scientific collaborations like the Internationale Gradmessung or the Carte du Ciel, which committed participants around the globe and across generations to instruments, procedures, and research agendas standardized in the name of commensurability and solidarity. Charles Sanders Peirce, who himself participated in some of these far-flung collaborations as an experimental physicist, drew the philosophical moral that scientific objectivity depended on the existence of a vast scientific community, extended over time and space, "beyond this geological epoch, beyond all bounds."[23] Or as the experimental physiologist Claude Bernard put it with lapidary concision: "L'art c'est moi, la science c'est nous."[24]

But if science—and with it, objectivity—had come to be identified with the communal and the communicable, how did art wander to the pole of solitude and the individual? Within the Enlightenment framework, both savants and artists, especially those touched by genius, were often idealized as solitary seekers of deep truths on the model of hermetic saints, whatever the biographical realities might have been.[25] One might therefore argue that there is nothing to be explained on the side of art: artists, at least in their idealized personae, simply remained lonely geniuses while their scientific brethren became clubby, and thereby ungenial. So simple a conclusion would, however, overlook the impact of far-reaching changes in aesthetics and in views of the artistic imagination that occurred in the early decades of the nineteenth century. Again, I can offer only a small sampling over the many possible examples to make my point vivid.

The ramifications of post-Kantian theories of the imagination fan out into a broad and branching tree, from Johann Gottlieb Fichte to Friedrich Schelling to Samuel Taylor Coleridge to Jules Michelet and beyond.[26] There is probably no generalization that holds for all of these developments, but tendencies can be discerned. For my purposes, the most significant are, first, the heavy emphasis upon the almost mystical orginality of the imagination, independent or even in defiance of reason and will; and, second, the allied cult of individual subjectivity, what the art historian Rudolf Wittkower once called the "egomania" of romanticism. Each element had distinguished antecedents—Plato's poetic furor or the Renaissance master as *Deus artifex*—but the combination of the two was novel to the early nineteenth century. Quasi-divine inspiration overwhelming will and judgment had not been traditionally paired with towering individualism: for example, the pythian priestesses through whom the godhead spoke at the oracle of Delphi were inspired but interchangeable. The intertwining of these two elements—originality and subjectivity—effectively rehabilitated what Enlightenment theorists had regarded as the pathological imagination. For the romantics, it was the unbidden, darkling force of the so-called passive imagination that was the wellspring of genial creativity, not the well-regulated active imagination subservient to will and reason. As William Blake retorted to Sir Joshua Reynolds, "What has Reasoning to do with the Art of Painting? . . . One power alone makes a poet; Imagination, the Divine Vision."[27] Hence the strong association in nineteenth-century psychological studies of genius—which restricted their subjects almost exclusively to artists and poets—between extraordinary creativity and the unconscious, or even insanity.[28]

In conjunction with the elevation of the passive imagination aesthetic norms shifted away from verisimilitude. A genuinely productive, as opposed to reproductive, imagination could be bound neither by the rules of decorum nor those of the natural order. Strict mimesis had never been the avowed ideal of Enlightenment critics, but they had subscribed to a standard of truth to nature, if not truth to fact. Romantic poets and artists attacked this aesthetic openly, under the twin banners of originality and individual subjectivity. Charles Baudelaire parodied what he called the credo of nature—"I believe in nature, and only in nature"—and called for art infused with imagination, for landscapes in which "human egotism replaces nature," for works to which the artist or poet "adds his soul." It was idolatry for art to prostrate itself before nature; any photograph could surpass the most faithful artistic replica in "absolute material exactitude." Deploring the public infatuation with photographic landscapes and portraits, Baudelaire insisted that ideals of truth and beauty not only did not coincide, they were inalterably opposed to one another: "With us the natural painter, like the natural poet, is almost a monster. The exclusive taste for the True (noble though it may be when limited to its true applications) here oppresses and suffocates the taste for the Beautiful." For Baudelaire, imitation of nature shaded imperceptibly into imitation of other artists: "The artist, the true artist, the true poet . . . must be really faithful to his own nature. He must avoid like death borrowing the eyes and the sentiments of another man, however great," just as he must avoid depicting "the universe without man," without the intervention of the imagination.[29]

It is customary to classify such views as "romantic," a term Baudelaire himself occasionally used. However, this label covers over fault lines that opened up within romanticism between subjective art and objective science, between the acolytes of beauty and those of truth. Although early nineteenth-century science had its own avowed romantics, such as Johann Wolfgang Goethe, Johann Ritter, Sir Humphrey Davy, or Alexander von

Humboldt, they were notably wary of the exalted imagination and individualism of the new aesthetics. The experimental physicist Ritter, who discovered ultraviolet radiation in his search for polarities in nature and who was given to utterances such as "Light is the external intuition of gravity, love the internal," nevertheless balked at allowing the imagination free rein in science: "The most beautiful thoughts are often no more than soap bubbles: filled with the hydrogen of our fantasy they rise quickly, and one does not realize that all the delightful play of their colors is nothing more than the reflection of their deceptive interiors."[30] Goethe warned the experimentalist against "the imagination [*Einbildungskraft*], which raises him to heights on its wings while he still believes his feet to be firmly planted on the ground";[31] Alexander von Humboldt scrupulously divided his monumental survey of nature into a first part containing "the main results of observation, which, stripped of all the extraneous charms of fancy, belong to the purely objective domain of a scientific delineation of nature," and a second part on "impressions reflected by the external senses on the feelings, and on the poetic imagination of mankind."[32] The wild imagination and individualism now held to be the birthright of true artists frightened even romantic scientists.

The point is that the newly erected divide between the objective and the subjective—the very words first enter dictionaries as a pair in German, French, and English in the 1820s and 1830s[33]—ran deeper than any opposition between neoclassicism and romanticism. My claim is not that there ceased to be fastidious realists among artists or daring speculators among scientists. Baudelaire found plenty of nature-worshipers to criticize among the paintings on display at the Paris Salon of 1859; Tyndall did not want for examples of scientists guided by their sense of beauty. But the new polarity of the objective and subjective structured how such boundary-straddling was perceived. When the novelist Gustav Flaubert attempted in *Madame Bovary* (1856) to depict a provincial adultery with clinical, impartial accuracy, both he and his critics seized upon the word "objective" to describe a style in which "subjects are seen as God sees them, in their true essence."[34] When embryologist Wilhelm His described the advantages of scientific drawings, he called the result "subjective."[35] Successful art could and did emulate scientific standards of truth to nature, and successful science could emulate artistic standards of imaginative beauty. But whereas in the eighteenth century both artists and scientists had seen no conflict in embracing both standards simultaneously, the chasm that had opened up between the categories of objectivity and subjectivity in the middle decades of the nineteenth century—words that, as Thomas De Quincey wrote in 1856, had once sounded pedantic and yet had so quickly become "indispensable to accurate thinking and to *wide* thinking"[36]—forced an either/or choice.

Hence a figure like Goethe, who combined artistic and scientific interests, became an uncomfortable paradox, especially for German scientists who could hardly escape the long shadow cast by the official national genius. The obligatory addresses delivered by leading German scientists on Goethe's scientific work provide a sensitive indicator of how entrenched the divide between objective and subjective had become. The physicist Hermann von Helmholtz gave two such addresses, in 1853 and 1862, and both turned on what Helmholtz took to be the opposition between scientific and artistic ways of thinking. Goethe's regrettable (in Helmholtz's view) attack on Newtonian optics could be explained, if not excused, by the impossibility of mingling the ineffable, almost divinatory intuitions of the artist with the crystalline concepts of the scientist. As in Kant's touchstone for distinguishing the objective from the subjective, communicability was central to Helmholtz's analysis of the distinction between artistic and scientific thinking: "Since artistic intuitions are not found by way of conceptual thinking, they cannot be defined in words. . . ."[37]

At the crossroads of the choice between objective and subjective modes stood the imagination. Very few nineteenth-century writers went so far as to deny scientists any imagination. Baudelaire, for example, acknowledged that imagination was as essential to the great scientist—or for that matter, the great diplomat or soldier—as to the artist. But in the next breath he relegated photography, whose exact rendering of what is seen he took to be diametrically opposed to the artistic imagination, to the sphere of science, where it might serve without corrupting.[38] By the last quarter of the nineteenth century, psychologists who investigated creativity routinely distinguished between different species of imagination, including the artistic and the scientific. In what was perhaps the most exhaustive treatment of the subject, the French psychologist Théodore Ribot defended science against the charges that it "sometimes extinguished the imagination" but nonetheless insisted that the "plastic imagination" of artists and poets and the "scientific" imagination belonged to different species (and further distinguished varieties within each species). Whereas the plastic imagination was free to invent and to grant its inventions a degree of emotional reality, the scientific imagination was constrained by "rational necessities that regulate the development of the creative faculty; it cannot wander aimlessly; in each case its end is determined, and, in order to exist, that is to say, in order to be accepted, the invention must be subjected to predetermined conditions."[39] For all his insistence on the existence and fecundity of the scientific imagination, Ribot could not free himself from a certain suspicion that imagination was linked to scientific error: the "false sciences" of astrology, alchemy, and magic represented for Ribot "the golden age of the creative imagination" in the history of science. In its 1902 survey of the psychology of creative mathematicians, the journal *Enseignement Mathématique* asked respondents, inter alia, whether "artistic, literary, musical, or, in particular, poetic occupations or relaxations seem to you of a nature to hinder mathematical invention, or to favor it, by the momentary rest they offer the mind?" It was apparently inconceivable that the exercise of the artistic imagination could promote the work of the mathematical imagination, except as a distraction in the same category as "physical exercises" and "vacations."[40]

CONCLUSION: ENDURING ART, EPHEMERAL SCIENCE

It is against this historical background that we must read distrust of the imagination in science. The power of the imagination had long awakened fear among scientists—and theologians, poets, artists, and doctors, to boot—because it could make up a world of its own that was livelier, lovelier, or more logical than the real world. In extreme cases the imagination could conquer the body as well as the mind, leading not only to madness but also to violent somatic crises. But Enlightenment theorists of the imagination had been confident in the right and competence of reason to discipline the imagination. Geniuses of art and science exercised the same brand of controlled imagination, in contrast to the wild imaginations that tyrannized pregnant women, religious fanatics, or mesmerized *convulsionnaires*. Only in the early nineteenth century was fear of the imagination in science compounded with loathing. The causes lay in new views of the artistic imagination as freed from all constraints of reason and nature, and in a new polarity between objectivity and subjectivity. Wild, ineffable imagination became the driving force of creativity in art—and the bogey of objectivity in science. In their ideals, practices, and personae both art and science had mutated, and drifted apart.

What kind of objectivity bans the imagination from science? I have mentioned one moment of objectivity, the communitarian impulse that urges scientists to standardize their instruments, clarify their concepts, and depersonalize their writing styles to achieve communicability and commensurability across continents and centuries, perhaps even across planets. Max Planck spoke in the name of this form of communitarian objectivity when he yearned for a physics that would be accessible "to physicists in all places, all times, all peoples, all cultures. Yes, the system of theoretical physics lays claim to validity not merely for the inhabitants of this earth, but also for the inhabitants of other heavenly bodies."[41] Communitarian objectivity could not coexist with the artistic cultivation of individualism, which enshrined personal perspectives and identified the ineffable with originality.

There was, however, a second moment of scientific objectivity that emerged alongside communitarian objectivity in the mid-nineteenth century. In an earlier article, Peter Galison and I have called this second moment "mechanical objectivity;" it replaces judgment with data-reduction techniques, observers with self-registering instruments, hand-drawn illustrations with photographs.[42] Mechanical objectivity strives to eliminate human intervention in the phenomena, to "let nature speak for itself." The free imagination celebrated by Baudelaire and other romantics threatened mechanical objectivity by projecting its own creations onto the facts of nature. Yet the facts envisioned by nineteenth-century scientists were not the fragile, pliable facts so carefully protected by their eighteenth-century predecessors from the distortions of system-builders. It was a byword that facts were angular, even truculent entities, sturdily resisting all attempts to ignore them or bend them to fit the Procrustean bed of theory. Huxley insisted that "a world of facts lies outside and beyond the world of words."[43] In part, this change in scientific perception corresponded to a very concrete change in scientific practice: in the last quarter of the eighteenth century a new generation of instruments and measuring techniques made it possible to stabilize and replicate results with a success undreamed of fifty years earlier.[44] In some real sense, scientific facts had become more robust. Why, then, were the automated ideals and practices of mechanical objectivity necessary at all? Why couldn't hard facts defend themselves against wild imagination?

The answer lies in a very different kind of fear that began to haunt scientists in the 1830s—the fear of vertiginous, open-ended progress. When Kant denied scientists genius, he had consoled them with progress: "The talent for science is formed for the continued advances of greater perfection in knowledge, with all its dependent practical advantages, as also for imparting the same to others. Hence scientists can boast a ground of considerable superiority over those who merit the honor of being called geniuses, since genius reaches a point at which art makes a halt, as there is a limit imposed upon it which it cannot transcend."[45] In the late eighteenth century, the sciences did indeed seem destined for smooth, steady, unlimited progress. Between 1750 and 1840, a stream of histories of various sciences poured from the press, all purporting to demonstrate the existence and extent of progress in those disciplines.[46] But the progress envisioned in these optimistic histories was of change without transformation. Once the foundations for the new science had been laid in the seventeenth century, as the standard story went, the edifice could be expanded but not remodeled. In the 1830s this placid view of scientific progress received a rude shock when the wave theory unseated the Newtonian emission theory of light, most notably as a result of the research of French physicist Augustin Fresnel.[47] How could a tested theory of impeccable

scientific credentials, its luster burnished by the name of Newton, be so thoroughly routed—not merely generalized or simplified? Was scientific progress so inexorable, so durable after all?

The response of scientists was to retreat to the level of the
description of facts, in order to salvage a stable core of knowledge from the ebb and flow of theories. As Ernst Mach put it in
1872, history of science taught the Heraclitean lesson of *panta
rhei*, for revolutions in science had become perpetual: "The
attempts to hold fast to the beautiful moment through textbooks
have always been futile. One gradually accustoms oneself [to the
fact] that science is incomplete, mutable."[48] Mach held up Joseph Fourier's heat theory as a "model theory" in science because it wasn't really a theory at all, being founded only on
"observable fact."[49] The expectations for scientific progress voiced
by Kant and others had not been disappointed; rather, they had
been fulfilled with a vengeance. Never before had science bustled
and flourished as it did in the latter half of the nineteenth
century. Scientists multiplied in number, and with them, new
theories, observations, and experiments. With these efforts, however, science not only grew; it also changed, and changed at a
rate that could be measured in months rather than generations.
No theory was safe from this breakneck progress, not even
Newtonian celestial mechanics. By the 1890s Henri Poincaré
was calling for ever more precise techniques of approximation in
order to test whether Newton's law alone could explain all
observed astronomical phenomena.[50]

Within this maelstrom of change, only facts seemed to hold
out the hope of definitive achievement in science. Like dia-
monds, scientific facts not only hardened but grew more pre-
cious to scientists in the nineteenth century—hence the fervor of
proponents of mechanical objectivity in fending off all possible
adulterations and distortions of facts by judgment or, especially,
imagination. Eighteenth-century savants had revered facts but
had believed them to be the alpha, not the omega, of scientific
achievement. (It should be noted that in eighteenth-century clas-
sifications of knowledge the custodians of fact were not natural
scientists, per se, but rather civil and natural historians.) Moreover, they were confident that facts mangled by the *esprit de
système* or an errant imagination would ultimately be corrected

by theory. Their nineteenth-century successors, caught up in the gallop of progress, had lost this innocent trust in the corrective power of theories that came and went like mayflies. Pure facts, severed from theory and sheltered from the imagination, were the last, best hope for permanence in scientific achievement. As anthropologists teach us, loathing stems from some breach of purity, some sacred boundary transgressed. The wild imagination potentially contaminated the purity of facts, and this is why it came not only to be feared but also loathed.

There is a rusting irony in the reversed fortunes of art and science, already visible in the mid-nineteenth-century writings of scientists. Alexander von Humboldt sadly reflected in 1844 on the contrast between ephemeral science and enduring literature, saying, "It has often been a discouraging consideration, that while purely literary products of the mind are rooted in the depth of feelings and creative imagination, all that is connected with empiricism and with fathoming of phenomena and physical law takes on a new aspect in a few decades, . . . so that, as one commonly says, outdated scientific writings fall into oblivion as [no longer] readable."[51] By 1917 Max Weber could regard the opposition of transitory science to stable art to be a platitude, one that made it difficult to understand what sense it made to pursue science as a career. Near the end of World War I, addressing an audience of Munich students who desperately wanted him to explain how science illuminated the meaning of life, Weber flatly asserted that science provided no such answers; science could hardly answer the question of what the meaning of a scientific career was. Why should one devote a lifetime of labor to producing a result that "in 10, 20, 50 years is outdated"? Subjective art endured, but objective science evaporated. Weber's own answer crowned this irony with yet one more. The spiritual motivation and reward for a lifetime devoted to science was exactly the same as for a lifetime devoted to art: science for science's sake, art for art's sake, the immolation of the personality in the service of "the pure object alone."[52] Having disavowed the artistic imagination and having lost the permanence of artistic achievement, science nonetheless aspired to the ascetic single-mindedness of art.

92   *Lorraine Daston*

ENDNOTES

[1] Aaron Fischbach, "In the Mail," *New Yorker* (22 July 1996), 6. Kevles's article appeared in the 27 May 1996 issue of the *New Yorker*.

[2] *London Times*, 19 September 1870; reprinted in John Tyndall, *Essays on the Use and Limit of the Imagination in Science* (London: Longmans, Green, & Co., 1870), 1–2.

[3] Gerald Holton, "Imagination in Science," in his *Einstein, History, and Other Passions: The Rebellion against Science at the End of the Twentieth Century* (Reading, Mass.: Addison-Wesley, 1996), 78–102.

[4] For an account of the transformation of scientific experience in the early modern period, see Peter Dear, *Discipline and Experience: The Mathematical Way in the Scientific Revolution* (Chicago: University of Chicago Press, 1995).

[5] René Descartes, *Discours de la méthode* [1637], Part VI. Unless otherwise specified, all translations are my own.

[6] Francis Bacon, *Novum organum* [1620], I.46.

[7] *Histoire de l'Académie Royale des Sciences. Année 1699*, 2nd ed. (Paris, 1718), "Préface."

[8] See relevant entries in the *Oxford English Dictionary*, *Grimms Wörterbuch*, and the *Dictionnaire historique de la langue française*.

[9] Bernard de Fontenelle, *De l'origine des fables* [1724], ed. J.-R. Carré (Paris: Librairie Félix Alcan, 1932), 14, 33.

[10] Samuel Johnson, *The History of Rasselas Prince of Abissinia* [1759], ed. J.P. Hardy (Oxford: Oxford University Press, 1968), 98–99, 104–105.

[11] René François Ferchault de Réaumur, *Mémoires pour servir à l'histoire des insectes*, 6 vols. (Paris: Imprimerie Royale, 1734–42), vol. 2 (1736), xxxiv–v.

[12] Georges Cuvier, *Recueil des éloges historiques lus dans les séances publiques de l'Institut de France* [1819–27], 3 vols. (Paris: Firmin Didot Frères, Fils, 1861), vol. 3, 180.

[13] Quoted in Rudolf Wittkower, "Genius: Individualism in Art and Artists," in Philip P. Wiener, ed., *Dictionary of the History of Ideas*, 4 vols. (New York: Charles Scribner's Sons, 1973), vol. 2, 297–312, here 307, 308.

[14] [Voltaire], "Imagination," in Jean d'Alembert and Denis Diderot, *Encyclopédie, ou Dictionnaire raisonné des sciences, des arts et des métiers*, 17 vols. (Paris/Neuchâtel, 1751–80), vol. 8, 560–563. See also the essay by Wendy Doniger and Gregory Spinner in this issue.

[15] [Marmontel], "Fiction," in ibid., vol. 6, 679–683.

[16] [Chevalier de Jaucourt], "Vraisemblance," in ibid., vol. 17, 484.

[17] Voltaire, "Imagination," 561.

18[J.S. Bailly], *Rapport des commissaires chargés par le Roi, de l'examen du magnétisme animal* (Paris: Imprimerie Royale, 1784), 18–19. On animal magnetism in late eighteenth-century France and the background to the Royal Commission, see Robert Darnton, *Mesmerism and the End of the Enlightenment in France* (Cambridge, Mass.: Harvard University Press, 1968).

19Bailly, *Rapport*, 59.

20Denis Diderot, "Génie," *Encyclopédie*, vol. 7, 581–584, here 583.

21Immanuel Kant, *The Critique of Judgment* [1790], trans. James Creed Meredith (Oxford: Oxford University Press, 1952), II.i.46–47, pp. 168–170.

22Immanuel Kant, *Critique of Pure Reason* [1781, 1787], trans. Norman Kemp Smith (New York: St. Martin's Press, 1965) A820–21/B848–49, p. 645.

23Charles Sanders Peirce, "Three Logical Sentiments," in Charles Hartshorne and Paul Weiss, eds., *Collected Papers of Charles Sanders Peirce*, 8 vols., (Cambridge, Mass.: Harvard University Press, 1932), vol. 2, 396–400, here 398.

24Claude Bernard, *Introduction à l'étude de la médecine expérimentale* [1865], ed. François Dagognet (Paris: Garnier-Flammarion, 1966), 77.

25Dorinda Outram, "The Language of Natural Power: The 'Eloges' of Georges Cuvier and the Public Language of Nineteenth-Century Science," *History of Science* 16 (1978): 153–178.

26For general overviews of these developments, see Karl Homann, "Zum Begriff Einbildungskraft nach Kant," *Archiv für Begriffgeschichte* 14 (1970): 266–302; Mary Warnock, *Imagination* (London: Faber and Faber, 1976); Eva T.H. Brann, *The World of the Imagination: Sum and Substance* (Savage, Md.: Rowman and Littlefield, 1991).

27Quoted in Wittkower, "Genius," 306.

28Most famously in Cesare Lombroso, *Genio e follia. Prelezione ai corsi di antropologia e clinica psichiatrica* (Milan: G. Chiusi, 1864).

29Charles Baudelaire, "Salon de 1846," "Salon de 1859," in *Curiosités esthétiques. L'art romantique et autres oeuvres critiques*, ed. Henri Lemaitre (Paris Éditions Garnier, 1962), 97–200, 305–396.

30Johann Ritter, *Fragmente aus dem Nachlasse eines jungen Physikers*, ed. Steffen and Brigit Dietzsch (Leipzig/Weimar: Gustav Kiepenhauer, 1984), 96, 260.

31Johann Wolfgang Goethe, "Der Versuch als Vermittler von Objekt und Subjekt," [1792, publ. 1823], in *Goethes Werke*, 14 vols., vol. 13 (*Naturwissenschaftliche Schriften I*, eds. Dorothea Kuhn und Rike Wankmüller) (München: C.H. Beck, 1994), 14–15.

32Alexander von Humboldt, *Cosmos* [1844], trans. E.C. Otté and W.S. Dallas, 5 vols. (New York: Harper and Brothers, 1850–59), vol. 2, 19.

[33]For a brief account of the history of the words, see Lorraine Daston, "How Probabilities Came to Be Objective and Subjective, " *Historia Mathematica* 21(1994): 330–344.

[34]Quoted in Erich Auerbach, *Mimesis. The Representation of Reality in Western Literature* [1946], trans. Willard R. Trask (Princeton: Princeton University Press, 1953), 487.

[35]Wilhelm His, *Anatomie menschlicher Embryonen* (Leipzig: F. C. W. Vogel, 1880), 6. I am grateful to Robert J. Richards for this reference.

[36]Thomas De Quincey, *The Confessions of an English Opium Eater* [1821], in *The Works of Thomas De Quincey*, 2nd ed., 15 vols. (Edinburgh: Adam and Charles Black, 1863), vol. 2, 265.

[37]Hermann von Helmholtz, "Über Goethes naturwissenschaftlichen Arbeiten,"[1853] and "Goethes Vorahnungen kommender naturwissenschaftlichen Ideen," in *Vorträge und Reden*, 4th ed., 2 vols. (Braunschweig: Friederich Viewig und Sohn, 1896), vol. 1, 25–47; vol. 2, 335–361, here 344.

[38]Baudelaire, "Salon de 1859," 319–322.

[39]Théodore Ribot, *Essai sur l'imagination créatrice* (Paris: Félix Alcan, 1900), 198.

[40]"Enquête sur la méthode de travail des mathématiciens," *Enseignement Mathématique* 4 (1902): 208–211, Questions 19, 26, 28. Jacques Hadamard, *The Psychology of Invention in the Mathematical Field* [1945] (New York: Dover, 1949), reports on some results of the survey.

[41]Max Planck, *Acht Vorlesungen über theoretische Physik* (Leipzig: S. Hirzel, 1910), 6.

[42]Lorraine Daston and Peter Galison, "The Image of Objectivity," *Representations* 40 (Fall 1992): 81–128.

[43]Thomas Henry Huxley, "Scientific Education: Notes of an After-Dinner Speech," [1869], *Science and Education. Essays* (New York: Appleton, 1894), 115.

[44]Christian Licoppe, *La formation de la pratique scientifique: Le discours de l'expérience en France et en Angleterre (1630–1820)* (Paris: Editions de la Découverte, 1996), 243–317.

[45]Kant, *Critique of Judgment*, 170.

[46]Rachel Laudan, "Histories of Sciences and Their Uses: A Review to 1913," *History of Science* 31 (1993): 1–34, especially 5–12.

[47]For a detailed account of this episode, see Jed Z. Buchwald, *The Rise of the Wave Theory of Light. Optical Theory and Experiment in the Early Nineteenth Century* (Chicago: University of Chicago Press, 1989).

[48]Ernst Mach, *Die Geschichte und die Wurzel des Satzes von der Erhaltung der Arbeit* [1872], 2nd ed. (Leipzig: Johann Ambrosius Barth, 1879), 1.

[49]Ernst Mach, *Die Principien der Wärmelehre* (Leipzig: Johann Ambrosius Barth, 1896), 115.

[50]Henri Poincaré, *Les Méthodes nouvelles de la mécanique céleste*, 3 vols. (Paris: Gauthier-Villars, 1892–99), vol. 1, 3–4.

[51]Humboldt, *Cosmos*, vol. 1, xxiv.

[52]Max Weber, "Wissenschaft als Beruf," [1917] in *Max Weber Gesamtausgabe*, ed. Wolfgang J. Mommsen and Wolfgang Schluchter with Birgitt Morgenbrod (Tübingen: J.C.B. Mohr, 1992), vol. 17, 84–87.

If there was synthesis anywhere, it must surely be sought in the mind of thinkers; there was no trace of it anywhere in practice. Toward the end of this life, Voltaire wrote a tale, *La princesse de Babylone,* in which he used once more the favorite eighteenth-century pattern of the intelligent and curious traveler, this time a pair of lovers, Amazan the Gangaride and Formosante the Princess, who pursue one another in a vast *chassé-croisé* across Europe and Asia; as they go a chapter is devoted to each country they visit, and the impression is finally conveyed that there is no unity or uniformity, no coherence or understanding in all this diversity of cultures, that the only order that can be found must surely lie in the tolerant outlook and liberal intelligence of the enlightened observer.

—Harcourt Brown

"Science and the Human Comedy: Voltaire,"
from *Dædalus* Winter 1958,
"Science and the Modern World View"

Wendy Doniger and Gregory Spinner

Misconceptions:
Female Imaginations and Male
Fantasies in Parental Imprinting

INTRODUCTION: SEND IN THE CLONES

PEOPLE USED TO JOKE that if a child was born with certain characteristics, it was because the mother, when pregnant, had been frightened by someone or something that had those characteristics. Some still cosset pregnant women to inculcate happy thoughts in them and to protect them from shocking or unpleasant thoughts; our reference to "strawberry marks" is probably an atavism of the belief that such marks reflect the pregnant woman's frustrated desire for strawberries. The folk view that was the prevalent view of the premodern world is still a part of the unofficial postmodern worldview, submerged in our unexamined habits of speech and custom. The man's desire to control the woman's desire, as it might affect his offspring, strongly colors our emotional reactions to abortion, the extreme case of a woman's desire to assert her agency over not merely the form but indeed the very life of the embryo.

A surprisingly large number of people, in different cultures over many centuries, have believed that a woman who imagines or sees someone other than her sexual partner at the moment of conception may imprint that image upon her child—thus prede-

Wendy Doniger is the Mircea Eliade Distinguished Service Professor of the History of Religions at the University of Chicago.

Gregory Spinner is Visiting Assistant Professor at Tulane University.

termining its appearance, aspects of its character, or both. This essay will consider a number of stories about the workings of maternal imagination, impression, or imprinting, terms that are often conflated. We will argue for a clear distinction between impression (the mental reception, and transmission to the embryo, of a visual image that is physically present) and imagination (a fantasy about something or someone who may not be physically present); together, we will refer to them as imprinting. And, since we will also consider the far less common (but equally relevant) instances of paternal imprinting, and since maternal imprinting itself only became problematic as it threatened the assumed paternal imprinting, it might be better to address the problem as *parental* imprinting.

Variants on the stories of parental imprinting may assume more or less the same mechanism of human embryology yet draw very different conclusions in different cultural contexts. The problem of the resemblance of a child to its parent(s) evoked the aesthetic question of the relationship between the original and the replicating image, as well as the theological question of the relationship between the activity of the Creator and the act of human procreation. By tracking the different stories and taking note of their distinctive features, we may reconstruct the lines of transmission within the traditions, suggest borrowings between traditions, and interrogate the shared premise.

The unexpressed assumption underlying most of these stories, and still a part of our own expectations, is that a male child should resemble his father ("chip off the old block") and, to some extent, his mother. The emphasis upon the male child reflects the androcentric concerns that drive most of our texts; the fact that some children do not resemble their parents excites anxieties about paternity and inheritance. As Thomas Laqueur remarks, "It is empirically true, and known to be so by almost all cultures, that the male is necessary for conception. It does not of course follow that the male contribution is thereby the more powerful one, and an immense amount of effort and anxiety had to go into 'proving' that this was the case."[1] The theory of parental imprinting was one way of accounting for divergences from the expected norm without admitting the likelihood of actual impregnation by an alienating male. This sort of mythological embryology involves a kind of pre-scientific cloning: it investigates ways of producing copies of desired stock. But, we must ask, desired by whom? One factor that seems to pervade all variants is the male desire to control female desire.

THE HEBREW BIBLE: JACOB'S PHALLIC RODS

Let us begin with a story from the Hebrew Scriptures. The patriarch Jacob promises to work for his father-in-law Laban, asking for his wages only the colored lambs and mottled kids from among the flock. This episode is recounted in Genesis, first in 30:25–43 and then in 31:1–12. In Genesis 31 the outcome of Jacob's wager with Laban is determined by God, as an angel reveals to Jacob in a dream; but the naturalistic explanation in Genesis 30 credits the clever use of ancient breeding techniques. Knowing that the specified mottling is unusual, Laban assumes that he will prosper from the deal, but this is not to be the case. Jacob takes fresh rods from almond, plane, and poplar trees and peels off strips and patches of their bark; he then places these variegated staves in front of the watering troughs. As the animals come to drink, they breed, and while they are breeding, the females stare at the rods. In this way, the patterns Jacob made by exposing the white of the wood are imprinted on the offspring; stripes, spots, and patches produce streaked, speckled, and brindled animals, respectively.[2]

The trick that Jacob played on Laban repaid Laban's trick on Jacob; as Laban had substituted Leah for Rachel (the object of desire) on the wedding night (Genesis 28:15–24), Jacob substituted variegated rods (phallic rods? ram-rods? the objects of desire) for the solid-colored rams within the field of vision of the ewes. The speckled ewes double for Rachel, whose name in Akkadian means "ewe"—a pun that plays a role in the scene in which Jacob meets, and desires, both Rachel and the sheep (Genesis 29:9–11; this is a conflation to which we will return in our consideration of later rabbinic texts). The biblical episode of the rods of Jacob became a paradigm often cited by later authors; by the process of prooftexting, and with the unsurpassed authority of Scripture, "the rods of Jacob" became a shorthand notation for the idea of maternal impression.

GREEK AND LATIN SOURCES: ARISTOTLE, EMPEDOCLES, SORAN, OPPIAN, HELIODORUS, AND JEROME

Aristotle remarked that the offspring of other animals resemble their parents more than human offspring do. He suggested that this might be because while animals are primarily concerned with the coupling, a human is not entirely filled with this desire but instead may be concerned with various things at the time of coupling, and the offspring become different from one another (*poikilletai*, "embroidered in different colors") in response to the concerns of the mother and the father.[3] A lost and probably apocryphal text attributed to Empedocles, a pre-Socratic poet (fl. fifth century B.C.E.) with whom Aristotle disagreed,[4] is quoted by Aetius: "How do offspring come to resemble others rather than their parents? [Empedocles says that] fetuses are shaped by the imagination of the woman around the time of conception. For often women have fallen in love with statues of men and with images and have produced offspring which resemble them."[5] The action begins with the mind, but the mental process quickly shifts to the eye, which passively receives the imprint of the artistic form (here, specifically an anthropomorphic form) and then, turning active, imprints that image upon the embryo. In keeping with the purely visual nature of the second stage of this replication, the child takes only external qualities from the mother's imagination. As with Jacob's ewes, the eye is the immediate organ of desire. But unlike Jacob, the husband in Empedocles's text plays no active role in supplying these artistic images; they may have been accidentally present or (dare one suggest?) actively procured by the wife. This is, as we shall see, a crucial difference.

The animal husbandry model in which the husband eugenically initiates the fantasy prevails in later Greek texts on this subject. In the *Gynecology* of Soran, an authority on obstetrics who lived at the turn of the second century C.E. in Rome and Alexandria, the husband plays the dual role of Jacob (masterminding the visual impressions) and the ram (impregnating the female):

> Some women, seeing monkeys during intercourse, have borne children resembling monkeys. The tyrant of the Cyprians, who was misshapen, compelled his wife to look at beautiful statues during intercourse and became the father of well-shaped children; and horse-breeders, during covering, place noble horses in front of the mares. Thus, in order that the offspring may not be rendered misshapen, women must be sober during coitus because in drunkenness the soul becomes the victim of strange fantasies; this furthermore, because the offspring bears some resemblance to the mother as well, not only in body but in soul. Therefore it is good that the offspring be made to resemble the soul when it is stable and not deranged by drunkenness.[6]

Soran assumes a correlation between human procreation and animal husbandry, comparing the tyrant who placed statues in front of his wife with horse-breeders who place handsome stallions (real ones, not images) in front of mares. He seems to have taken the folk wisdom recorded by Empedocles, that women *do* fall in love with statues, and connected it with the folk wisdom of animal husbandry recorded in Genesis (and elsewhere), that females can be *made* to desire obstetrically, as it were, the images that the husband desires eugenically; in the process, he has moved from the herd animals favored in the Bible (sheep and goats) to horses, the favorite animals of the Greeks. The result is an active attempt by the husband to treat his wife like a mare (or a ewe): he shows her images of what he wants her to give birth to. Soran pries into the psychology of a man who would do this: such a man might be ashamed of his own distorted form, and his fear that the child will not resemble the father (and will thus be illegitimate) is outweighed by his desire for a handsome heir.

The (human) males are in control in both halves of Soran's central episode, but it is framed by two others in which human men have no control at all. The stark, mindless physicality of the husbandry model has already been undercut by the first animals that Soran imagines the wife seeing—monkeys, far closer to the human than horses are and not so closely manipulated by humans in their breeding. This may account for the unusual (though still both limited and pejorative) agency granted to the woman's soul in the final episode, in which Soran considers distortions that arise not only in the mother's field of vision but in the inner vision of her imagination (though still excited by an external force, wine). He apparently assumes that both men and women have in their souls the spiritual quality that animates matter and makes parental imprinting possible. But even his acknowledgment that the offspring resemble the mother leads Soran quickly to emphasize the negative aspect of that maternal influence, the fear that the soul of a woman out of control—here not with lust, but with drunkenness—might, like the mis-shapen form of the father, make the child misshapen.

Where horse breeding serves Soran merely as an illustration of what he is really interested in, it is the central topic of the *Kynegetika*, attributed to Oppian, a Syrian of the late second or early third century C.E. Oppian veers from his line, horses and hounds, just long enough to apply the principles of their breeding to humans. First he describes subtle devices for "inscribing the foal while yet in his mother's womb": when "the mating impulse seizes the mare," the stallion is adorned with "spots of color" and brought to the mare like a bridegroom entering a bridal chamber; then, "the mare conceives and bears a many-patterned foal, having received in her womb the fertile seed of her spouse, but in her eye his many-colored form."[7] Then we come to the human species. The Laconians, we are told, place before their pregnant wives images of ancient demigods noted for their beauty (Narcissus and Hyacinthus, Castor and Polydeuces), as well as the gods Phoebus and Dionysus. The women look at these beautiful forms and, excited by their beauty, bear beautiful sons.

Again we have husbands—now normal men, not misshapen tyrants—actively encouraging their wives (treated like mares, an old Greek and Indian habit)[8] to give birth to children who do not resemble their fathers. But something theological has been added, which persists in later texts: the idea of the image of a celestial being imprinted on a human child. Here the divine is the model not of goodness or wisdom but simply of the external quality of beauty, as befits the simply visual mechanism of reproduction in this text.

Oppian talks about changing the color of animals and the beauty of human children, and later texts combined these ideas to produce the agenda of changing the color of human children. Halfway through the *Ethiopica* of Heliodorus, who lived in Syria in the third or fourth century C.E., we learn that on the armband worn by the heroine Charikleia there is an inscription from her mother, Persinna, queen of the Ethiopians, explaining why she had abandoned her child. The part of the inscription that concerns us reads:

> Our line descends from the Sun and Dionysos among gods and from Perseus and Andromeda and from Memnon too among heroes. Those who in the course of time came to build the royal palace . . . made use of the romance of Perseus and Andromeda to adorn the bedchambers. It was there one day that your father and I happened to be taking a siesta in the drowsy heat of summer. . . . Your father made love to me, swearing that he was commanded to do so in a dream, and I knew instantly that the act of love had made me pregnant. . . . But you, the child I bore, had a skin of gleaming white, something quite foreign to Ethiopians. I knew the reason: during your father's intimacy with me the painting had presented me with the image of Andromeda, who was depicted stark naked, for Perseus was in the very act of releasing her from the rocks, and had unfortunately shaped the embryo to her exact likeness. I was convinced that your color would lead to my being accused of adultery, for what had happened was so fantastic that no one would believe my explanation. . . .[9]

Andromeda was the daughter of the king of Ethiopia, but Greek artistic convention generally represented her with white skin.[10] In the seventeenth century, Fortunio Liceti objected to this aspect of the story: "To a natural philosopher's eyes, since Andromeda was born to Cepheus and Cassiopeia, the king and queen of Ethiopia, she was black."[11] But this is precisely why Andromeda is invoked here: though she was regarded as racially black, she was conventionally represented as white. She is thus the ideal liminal creature to lure Persinna across the line. And so Charikleia, whose lack of resemblance to her own parents is problematic, resembles her ancestor Andromeda in three ways: she is the daughter of an Ethiopian king, she has a romance with the hero Theagenes (like Andromeda with Perseus), and she has white skin. By emphasizing the color rather than the form or beauty of the child, Heliodorus is drawing upon the literature of animal husbandry, which emphasizes unusual color. But now this color is associated with a race of people, the Ethiopians; hence it is a racial, if not necessarily racist, story.[12] Certainly it provoked racist reactions in Europe; the editors of the French edition of Soran compared his Cyprian tyrant with Heliodorus's Ethiopian queen and blithely remarked, "One must not forget that, for a young black woman like the princess of the *Ethiopica*, the most beautiful baby in the world is a white baby."[13]

Heliodorus, like Soran, tells of a woman who sees an image that her husband has not actively intended her to see. As a result, the wife is afraid that she might be accused of adultery; but since she knows she is innocent, she deduces the cause of the lack of resemblance and preempts any accusation by abandoning the baby. In fact, her fears prove well justified; when Charikleia, years later, claims her heritage, her father, King Hydaspes, insists that she cannot be his child: "Your skin has a radiant whiteness quite foreign to Ethiopian women. . . . How could we, Ethiopians both, produce, contrary to all probability, a white daughter?"[14] For proof, the painting of Andromeda is brought out of the bedroom, and Charikleia stands beside it: "The exactitude of the likeness struck them with delighted astonishment." But the final proof of her identity, in addition to the cultural ring that her mother had left her (the ring that Hydaspes had given her at their wedding), is a natural ring, her birthmark, "like a ring of ebony staining the ivory of her arm!" Thus she is black after all, at least in that mark from her mother that answers to the ring of patriarchy.

Persinna calls this sort of maternal impression unbelievable, but many people were ready to believe it. Indeed, this very story of the black queen with the white daughter was retold on countless occasions. Jerome, one of the church fathers, writing between 386 and 390 C.E., regards it as "not astonishing" and uses it to gloss the story of Jacob and the rods. Then he remarks:

Now it is not astonishing that this is the nature of female creatures in the act of conception: the offspring they produce are of such a kind as the things they observe or perceive in their minds in the most intense heat of sexual pleasure. For this very thing is reported by the Spaniards to happen even among herds of horses; and Quintilian, in that lawsuit in which a married woman was accused of having given birth to an Ethiopian, brought as evidence in her defense that what we have been describing above is a natural process in the conception of offspring.[15]

Jerome, like the Greek writers, then moves from sheep to horses and goes on to cite a lost *controversia* (a legal fiction or hypothetical case) by Quintilian, from the first century C.E., involving Ethiopians. Though Jerome does not tell us the significance of the Ethiopian child of presumably non-Ethiopian parents, we might assume that, linked as it is with Jacob's rods, it has something to do with color.

Quintilian's colors—a white woman giving birth to a black child—represent a more logical choice of a meaningful problem for a white author, racist or not. However, it was Heliodorus's version (the black woman giving birth to a white child), not Quintilian's, that was more often cited in later Jewish and Greek literature.[16] No matter which way the colors flow, the underlying assumption remains the same: what a woman looks at when she is pregnant or at the moment of conception influences the physical nature, including the color, of the child.

The idea of maternal impression through artistic influence remained a part of the medical tradition in medieval Europe. Maimonides, the celebrated Jewish physician and philosopher, writing in Arabic at the end of the twelfth century C.E., took the theme of the imitation of a painting from the Greek medical tradition:

> I heard from the ancient physicians that he who wishes to give rise to a handsome son should request a very famous painter to prepare a portrait having the likeness of a beautiful child. He should then request of his wife that, during intercourse, she look at the portrait without winking and not move her eyes right or left. And so it happened that she gave birth to a beautiful child who resembled the portrait of the painter and did not resemble his father at all.[17]

Here, as in Soran and Oppian, the woman is given the image by her husband. Yet now it is an image of the desired child, not a handsome man who might perhaps incite the husband's jealousy; and the woman must stare without blinking or glancing aside, a daunting prospect in even the most abbreviated act of intercourse, but also an extreme form of the emphasis on the physical process of vision.

LATER JUDAISM:
VARIOUS MIDRASHIM, THE HOLY EPISTLE, AND THE ZOHAR

The child of a different color than its parents is a leitmotif of later texts that continue to transfer the embryological principle from animal husbandry to human eugenics. One of the oldest compilations of Jewish commentaries, roughly contemporaneous with Heliodorus, presents several glosses on the verse, "And Jacob took the rods" (Genesis 30:37). The first two glosses, like Genesis 31, explain the episode of Jacob and Laban's flock by miracles; the water in the troughs miraculously turned to semen, says one, while the other suggests that angels came down to help Jacob. But the third gloss, like Jerome, connects this text with the Ethiopian tale—though it uses Heliodorus's version of the colors, not Quintilian's:

> It so happened that a Kushite [Ethiopian] man married a Kushite woman who bore him a white son. The king seized the son and went to Rabbi. He said to him, "Consider whether he is my son or not." The other responded, "Are there pictures in your house?" "Yes." "Black or white?" "White." "Because of this, you have a white son."[18]

This version reads like a Cliffs Notes version of the story of Persinna giving birth to Charikleia.[19] Unlike Persinna, however, the Kushite woman does not realize that she will be accused of adultery, nor does she understand why she has a white child; she is robbed of the agency that she had in the Greek text. This Hebrew text, like many others, shifts its perspective from the mother to the reaction of the father, who is alarmed to have a white son; the unexpected change in skin color raises the suspicion of adultery. It would be hard to hazard a guess as to who (Quintilian, Jerome, Heliodorus, or the rabbis) got the story from whom (one from another, or all from another source). But it is possible that the rabbis found Heliodorus's story so apt an illustration of the idea of maternal impression, already present in Genesis 30, that they appropriated the episode, reworking it to their own ends.

The tale of the Ethiopian queen who gave birth to a white baby reappears in the rabbinic discussions of the judicial ordeal of the suspected adulteress:[20]

> Our rabbis said: When a woman is with her husband and is engaged in intercourse with him, and at the same time her heart is with another man whom she has seen on the road, there is no greater adultery than this; for it is said, "The wife commits adultery, taking strangers while under her husband" (Ezekiel 16:32). Can there be a woman who commits adultery *while under her husband*? It is this one, who has met another man and set her eyes upon him, and while she carries on intercourse with her husband, her heart is with him. The king of the Arabs put this question to R. Akiba: "I am black and my wife is black, yet she gave birth to a white son. Shall I kill her for having played the harlot while lying with me?" Said the other, "Are the figures in your house painted black or white?" "White," he said. The other assured him, "When you had intercourse with her, she fixed her eyes upon the white figures and bore a child like them. If you are surprised at such a possibility, study the case of our father Jacob's flock, which were influenced in their conception by the rods, as it says, 'And the flocks conceived at the sight of the rods'" (Genesis 30:39). The king of the Arabs acknowledged the justice of R. Akiba's argument. In our case as well, Moses hinted in the Torah at a similar situation by saying, "[If you have gone astray, though you are] under your husband, and if you be defiled, and some man has lain with you besides your husband . . ." (Numbers 5:20).[21]

Now the parents are Arab, but the racial point is the same (the version in *Tanhuma* says that the king of the Arabs told R. Akiba that he and his wife were Kushite, i.e., Ethiopian, but had a white child).[22] As usual, the prooftext is from Genesis 30, "the rods of Jacob," but now one of the characters cites the rods of Jacob to explain the Ethiopian problem, whereas the previous midrash told that Ethiopian story to explain the Biblical story. In fact, this text adduces both stories to explain the relationship between maternal imprinting and adultery, the latter epitomized in a verse from Ezekiel that is not about a human woman at all:

Jerusalem is personified as the unfaithful bride of the Lord. The biblical verse about adultery is directly linked with maternal imprinting by the midrashic process; Numbers 5:20 and Ezekiel 16:32 are intratextually correlated because the same phrase— "under her husband" (*tachath ishah*)—occurs in both verses. This phrase in Numbers is not read conventionally, meaning "under his control,"[23] but literally, meaning "under him physically," suggesting that the suspected adulteress is in fact embracing her husband while thinking of another man. The forceful pronouncement that "there is no greater adultery than this" indicates the high degree of anxiety aroused by the idea of female fantasizing, an anxiety that would be reinforced by an embryological principle empowering the maternal imagination to shape the child, distorting paternal resemblance. Thus questions are raised: Is the child still his child? Who is the true father?

Some fantastic considerations are taken up in the Babylonian Talmud in its discussion of the proper conditions and manner for conjugal intercourse, undertaken as a spiritual exercise. The Talmud cites a tale about Imma Shalom, wife of R. Eliezer, who was asked why her children were so beautiful; she credited her husband's pious conduct. More specifically, Imma Shalom had asked her husband why they engaged in conjugal relations only at midnight, and he had replied, "So that I do not set my eyes on another woman, begetting sons who are as bastards."[24] Such eugenic precaution assumes that in the middle of the night there would be no woman out and about upon whom Rabbi Eliezer might look and thus receive an image that he might then imprint upon his own progeny. This is a relatively rare example of paternal imprinting; Eliezer is concerned not that his wife will hear or see someone other than him, but that he will see or hear someone other than her. Moreover, it is a negative form of paternal imprinting: Eliezer makes sure that he does *not* make a false impression upon his unborn child.

Another story from the Babylonian Talmud, about Rabbi Yohanan, a sage renowned for his beauty, seems to contradict the story of Imma Shalom and Rabbi Eliezer:

> R. Yohanan used to go and sit at the gates of the *mikveh* [ritual bathhouse]. "When the daughters of Israel ascend from the bath,"

he said, "let them gaze upon me, so that they bear sons as beautiful and learned as I."[25]

The *mikveh* is where Jewish women go to prepare themselves for the sabbath, the last public place they would visit before returning home. As it is customary to engage in intercourse on the Sabbath eve, the beautiful image of R. Yohanan would still be freshly impressed on their minds' eyes when they made love with their husbands. R. Yohanan was trying to do precisely what R. Eliezer was trying to prevent: the women visually impregnated by R. Yohanan, through a (not so) chance encounter on the street, would give birth to near-bastards. Subsequent wordplay on "the gaze" in this text suggests that many listeners or readers would see it as a tongue-in-cheek parody of Greco-Roman eugenic techniques; but there may also be an underlying sense of anxiety over the reproduction not of male children but of male cultural values, notably Torah study.[26] The faithful reproduction of ideas, as well as children, requires a certain family resemblance; in recreating the sage's teaching, one replicates his image.

A condensed version of the Ethiopian story appears in *The Holy Epistle (Iggeret HaKodesh)*, a thirteenth century Kabbalistic sex manual that elaborates upon motifs in the Babylonian Talmud. As usual, it cites the story of Jacob's rods as a prooftext, but it arranges the colors in the Quintilian fashion:

> [A queen] had a black baby though the king and she were white and extremely comely. The king wanted to kill her until a wise man came and said, "Perhaps you thought of a black man at the time of intercourse." They examined the matter and found black designs on the drapes in their conjugal room. She said that she had looked at these black figures during intercourse and thought of them. This is just like the sticks of Jacob.[27]

Of course, it isn't "just like" the sticks (the translator's terms for what we have been calling the rods) of Jacob: the imprinting is accidental, and problematic to the father, in part because the passive, accidental visual impression is here combined with an active, mental act on the part of the woman. But when the wife falls in love with art, she is innocent; loving the art is not adulterous.

According to this text, the wife fantasizes about her husband, who contemplates the archetypal *sefirot* (divine emanations constituting the fullness and mystical shape of the Godhead) as a cognitive template to stamp the child with the *imago dei*. The mother's impression here is negative, like Eliezer's: her role is to prevent adultery of thought, while the husband's impression is positive. Thus the paternity is twofold: the wife focuses her mind on her husband, so that the child physically resembles his biological father, while her husband focuses his thoughts on the supernal form, so that the child metaphysically resembles his Father in heaven.[28]

Theological considerations are at the heart of the *Zohar*'s commentary on Jacob's first child, Reuben, begotten upon Leah apparently on the wedding night, when Leah was substituted for Rachel:

> On the night when he had intercourse with Leah he was thinking of Rachel. He lay with Leah but thought of Rachel, and his semen followed his thought, but it was not intentional, for he did not know. . . . And because the Holy One, blessed be He, knew that it was not intentional and that Jacob had truthful thoughts during his desire, [Reuben] was not disqualified from being counted among the holy tribes. Otherwise he would have been disqualified.[29]

This is an instance of paternal rather than maternal imprinting, and it is theologized by supplementing the power of the father's intentions with the power of the Father's intentions. The translator explains that since Jacob had intercourse with Leah under the genuine impression that she was Rachel, his sin of adultery was not intentional; and since Reuben's conception took place when Jacob was thinking of Rachel, she was finally the channel through which the birthright, bypassing Leah's son Reuben, was transmitted to her own later-born son, Joseph, and thence to Joseph's sons. The meditations on the true mother of Reuben, Rachel or Leah, depend upon the fantasizing of Jacob and God.

Jacob's imagination transformed not the form but the status of his son, indeed of both of his sons, Joseph and Reuben. The Babylonian Talmud enumerates ten kinds of children who are like bastards but who are not legally recognized as such, including the "children of substitution" (*b'nei temurah*), born when either one of the parents was thinking of someone else.[30] Similarly, in the *Zohar*, a man who makes love to his wife while he is thinking about another woman is said to "sow false seed," and the child is considered a kind of changeling, particularly susceptible to evil influences.[31] While adultery of thought is legally distinct from the corporeal act of adultery,[32] it nonetheless produces near-bastards and is considered morally reproachable:

> If a man defiles himself by evil thoughts when he comes to have intercourse with his wife, and sets his thoughts and desires upon another woman, and emits semen with these evil thoughts, then his thought effects changes in the world below [i.e., an exchange of women in one's thoughts]. . . . The body of the child that he begets is called "a changeling" [because the body was created while the father "changed" his thoughts during procreation].[33]

This idea underlies the *Zohar* text that insists that Reuben was *not* a "changeling," because Jacob was thinking about the woman he should have been in bed with, and thought he was in bed with; this is why he had "truthful thoughts." Had he known he was in bed with Leah and still thought of Rachel, the child would have been a changeling.

CHRISTIAN EUROPE:
PARACELSUS, GOETHE, HOFFMANN, AND SCHNITZLER

Christians, too, tended to fantasize that their wives were fantasizing. That children usually resemble their parents remained the standard European folk opinion, despite dissenting voices of people such as Malebranche (and several authors before him), who "suggested that no two faces in all the world are absolutely identical, and that nature tolerates great diversity,"[34] and Jacques-André Millot, who argued (in Paris, in 1800), that "resemblance is uncanny. Contrary to the Aristotelian definition, he who does not resemble his parents is not a monster but a normal child; the monster is a rarity, the result of pure chance, he who *does* perfectly resemble his parents."[35]

But children who did not resemble their parents were, *pace* Malebranche and Millot, generally regarded as monstrosities that had to be explained by the theory of parental imprinting, which made an important distinction between external and internal objects of desire. Paracelsus, writing in Germany in the sixteenth century, argued that the man's fantasy was the source of his semen, and added, "Thus God has put semen into the imagination of man."[36] He granted that women's imaginations, too, could affect the embryo—yet in positive ways:

> Through the power of the imagination, women in such moments imagine a learned wise man, such as Plato or Aristotle, or a warrior, Julius or Barbarossa, or a great artist, like the painter Dürer . . . and so they will bear children like them. And there must be not just lust and desire, but also experience of these arts and wisdoms, in the same way as there is an experience when they see a fish. . . . Thus a woman hears an artist like a musician, or even a learned man, and has a desire for that, and gives the impression to the child: and even if she does not understand it, and cannot, nevertheless the child will show the effect.[37]

Paracelsus attributes to the woman's imagination the same positive effect that other texts assign to the woman's experience of looking at a painting (which, presumably, her husband has given her). What is unusual here is that mental qualities as well as physical are seen as transmitted in this way; stories about parental imprinting generally speak only of superficial resemblances. It is also unusual for the woman's accidental encounters with other men, artists and musicians, let alone her imagination of other men, like Plato or Aristotle, to produce benefits not planned but welcomed by her husband.

The positive eugenic benefits imagined by Paracelsus were, however, generally outweighed by the fear of the negative results of women's imaginations. Fortunio Liceti, in 1616, maintained that "though the father's imagination can affect him during the sexual act, the woman's is always at work, after copulation and during conception, when the fetus is formed."[38] Ah, but what were those women thinking about? Their husbands, Ambrose Paré had hopefully suggested in 1585: "One more commonly sees children who resemble their father than their mother because of the mother's great ardor and imagination during carnal copulation! So much so that the child takes on the form and the color of what she knows and imagines so strongly in her mind."[39] Now, the logic of this statement seems to imply that the woman normally imagines her husband and perhaps also looks at him, since the child normally resembles him. But two hundred years later, in 1788, Benjamin Bablot argued from the same premise (that women are more passionate and imaginative than men) to reach the opposite conclusion: "As a child presents sometimes more his mother's features than those of his father, those attributing the cause to imagination say that the mother's thoughts were completely absorbed by her loving passion during conception and were unable to focus on her husband's features."[40] And so, we may conclude, she focused on her own image, or even—heaven forbid!—her own pleasure. This line of reasoning had been made explicit in a medical case, cited by the Chevalier Sir Kenelm Digby in 1678, of a woman who kept gazing at her artificial beauty marks in a mirror and who gave birth to a child with such marks: "Instead of thinking about her husband, the mother has given in to . . . a narcissistic delight in her own image."[41] The possibility that she might have been thinking about another man, however, inspired still more devastating speculations.

Voltaire, writing in 1765, believed in the power of parental imprinting, despite himself: "This passive imagination of easily shaken brains often produces in children the visible marks of an impression that the mother has received; there are innumerable examples, and the present writer has seen such striking ones that he would accuse his eyes of lying if he doubted them, and although this influence of the imagination is inexplicable, no other influence explains the matter any better."[42] Montaigne did not doubt what he had heard: "There was presented to Charles, King of Bohemia and Emperor, a girl from near Pisa, all hairy and bristly, who her mother said had been thus conceived because of a picture of Saint John the Baptist hanging by her bed."[43] This could happen because the woman misinterprets what she sees, and her *mistaken* perception of what she has seen imprints the child: "Thus, the furs covering John the Baptist are 'translated' into a hair-covered body."[44] In the eighteenth century, the Siamese twins Judith and Helena were thought to be connected as they were "because early in her pregnancy their mother had been foolish enough to watch dogs mating."[45] We may also see in these texts atavisms of the ancient connection between maternal imprinting and animal husbandry. The theme remained popular in European fiction;[46] in James Joyce's *A*

Portrait of the Artist as a Young Man (1916), Stephen Dedalus imagines "the sleek lives of the patricians of Ireland" and wonders, "How could he hit their conscience or how cast his shadow over the imagination of their daughters, before their squires begat upon them, that they might breed a race less ignoble than their own?"[47] Like R. Yohanan, Stephen longs to lend his superior qualities to other men's wives, using not his body but his shadow—that is to say, his image, like the image in a painting.

The theme was particularly influential in Germany. In Goethe's novella *Elective Affinities* (*Wahlverwandtschaften*, 1809), *both* parents fantasize, and the child resembles a combination of the two fantasized lovers. On what amounts to their wedding night, Charlotte and Edward make love, but Charlotte is aware of the ghostly presence in the bedroom of the Captain, whom she really loves, and Edward, similarly seduced by the darkness and his own imagination, feels that he holds Ottilie in his arms. As a result, the son of Charlotte and Edward is the striking image of their secret loves:

> People saw in it a wonderful, indeed a miraculous child. . . . What surprised them more . . . [was] the double resemblance, which became more and more conspicuous. In figure and in the features of the face, it was like the Captain; the eyes every day it was less easy to distinguish from the eyes of Ottilie.[48]

The child—who thus reveals the effects of his parents' imaginations and betrays their moral adultery—dies in early childhood.

In E. T. A. Hoffmann's "The Doubles" ("*Die Doppelgänger*," 1821), two babies born to two different women look exactly alike; they completely resemble the man that only one of the women made love with, though the other loved him too: "Even if this could be chance or an illusion, the quite superior formation of the skull and a small moon-shaped mole on the left temple affirmed the complete similarity."[49] The father of the child that resembles the other man banishes mother and child. His jealousy is justified not physically but spiritually; the woman accuses herself of having committed adultery only in her mind but regards this as an inexpiable sin, a mental infidelity sufficient to cause the maternal imprinting. Her behavior outweighs mere visual considerations, but that behavior is a clue to her all-important mental state, which is ambivalent. People who were "closely acquainted" with her knew she could not have had an illicit affair, despite the visual evidence to the contrary, because she was so good; her husband, however, was swayed not by the visual evidence but by his wife's behavior—she hated her child.

The racial (or racist) aspects of maternal imprinting also continued to be expressed in European literature. Ambrose Paré, writing in the sixteenth century, retells the Heliodorus version of the Ethiopian queen and then tells the Quintilian version of the colors, for good measure:

> Hippocrates saved a princess accused of adultery, because she had given birth to a child as black as a Moor, her husband and she both having white skin; which woman was absolved upon Hippocrates' persuasion that it was [caused by] the portrait of a Moor, similar to the child, which was customarily attached to her bed.[50]

The story about Hippocrates was widely cited during the Renaissance.[51]

Arthur Schnitzler mocked the theory of maternal imprinting in his story, "Andreas Thameyer's Last Letter" (1918), about a suicide note written by a man unwilling to admit to himself that his wife had betrayed him with a black man:

> I've read the case described by Malebranche. And Martin Luther himself—as one can read in his after-dinner speeches—knew, in Wittenberg, a man who had a death-head because his mother had been frightened by the sight of a corpse while she was pregnant with him. [Here he cites Heliodorus's tale of the Ethiopian queen, and other cases noted by Hamberg, Weisenburg, Preuss, Limböck, and others.] In 1737 in France a woman gave birth to a son when her husband had been absent for four years, and she swore that she had, during that period, dreamt of the passionate embrace of her husband. The physicians and midwives of Montpellier declared that this was quite possible, and the court at Havre declared the child legitimate. . . .
>
> And it happened to me, and to my wife, who was true to me, as truly as I am living at this moment. Our child is now fourteen days old. . . . My wife has been true to me, and the child that she bore to me is my child. She had a shock in her pregnancy, in August, when she was at the zoo with her sister Fritzi, where these foreign people had camped, these uncanny black people

[*diese unheimlichen Schwarzen*]. . . . Wednesday she and her sister Fritzi went to the zoo, where Negroes had camped. I myself saw these people later, in September. . . . My wife was terrified, and alone, for Fritzi had suddenly left her to go off with a married man who has a rather bad reputation. . . . My wife waited for Fritzi for two hours, and then the gates were shut, and she had to go. She told me all this, with her arms around my neck as I sat on her bed, and she trembled in fear, and I was afraid, too, though I didn't know then that she was already carrying our child.[52]

The argument turns upon the layered meanings of the phrase, "*Sie hat sich versehen,*" which means, literally, "She mis-saw" (on the analogy of *verhören*, "to mis-hear"), that is, "She made an oversight, a mistake," then, "She had a (visual) shock," and more particularly, said of pregnant women, "She had a (visual) shock in her pregnancy," that is, "She received a maternal impression." (Havelock Ellis refers to "the conception of a 'maternal impression' [the German *versehen*]."[53]) So a pregnant woman who sees something mistakenly has a shock which imprints a mistake upon the embryo. In Schnitzler's story, the husband's suspicions of his wife are projected onto the disreputable married man with whom the wife's *sister* has an affair and are further inflamed by his racist attitude to the black people who shocked his wife. These emotions keep breaking through his insistence that this sort of thing can be explained by men of science, and they finally drive him mad.

The Christian theories grow out of the father's fear that his child may not be his child, or, rather, that he can never be sure that his child is his child—unless, of course, he trusts his wife. Resemblance was the straw that men grasped in the storm of their sexual paranoia. Montaigne cites Aristotle as saying that "in a certain nation where the women were in common they assigned children to their fathers by resemblance."[54] The fear that women might be, indeed, "in common," is what underlies this entire corpus, and various ideas about resemblance are conjured up in the attempt to lay that fear to rest.

In premodern Europe, the mother's imagination and desire gave birth to a "false resemblance" and were defined as illegitimate when they were not directed toward her husband.[55] Thus, as Marie-Hélène Huet has pointed out, nonresembling children served as "a public reminder that, short of relying on visible resemblance, paternity could never be proven." The monster, the child that did not resemble the father, unmasked what the theory of resemblance concealed, the fear of adultery; it "erased paternity and proclaimed the dangerous power of the female imagination."[56]

As if this were not bad enough, the theory was turned on its head to show that resemblance, too, could dissemble. Thus Nicolas Venette, in 1687, agrees with the wise lawyers and doctors who "claim that a woman who thinks strongly about her husband in the midst of illicit pleasures can produce, through the force of her imagination, a child that *perfectly resembles him who is not the father.... Resemblance is not proof of filiation....*"[57] Now, it is easy enough to imagine that a woman might dream of her lover while in the embrace of her husband, but why, one might ask, would she think of her husband while in the embrace of her lover? One answer is that if she, too, subscribes to the theory of maternal imprinting, she will think about her husband when she is with her lover on purpose, in order to conceal her adultery.[58] Thus, Huet argues, the mother is no longer regarded as a victim of her own passion or desire, but, rather, in control of her own imagination to such a degree that she can produce, through that imagination, "not a monster but its exact opposite: a child who actually resembles the legitimate spouse who did not father it."[59]

The backlash from this development was very serious. For where the theory of maternal imprinting doubtless saved the necks of a number of adulteresses whose children did not resemble their fathers, the corollary (that an adulteress could imprint her lover's child with her husband's features) cast suspicion upon *all* women, indeed particularly upon faithful women, whose children *did* resemble their legitimate fathers.[60] Women as a whole were portrayed as bodysnatchers, who could at will replace a seemingly normal child with a monster conceived in the pods of adulterous beds. This was truly a no-win situation; women were damned if they did, and damned if they didn't, produce children who resembled their husbands. The cognitive dissonance that resulted from this uncertainty drove men both to attempt to control women's sexuality and to project their images upon their sons. Thus M. Boursicot (the French diplomat

in the real-life affair that inspired David Henry Hwang's
M. Butterfly) explained why he thought his (male) Chinese lover
was a woman who had borne him a son by saying of the child,
"He looked like me."[61]

There are also counterinstances in which the woman's revulsion from a man or his image, rather than her desire for him, produces another sort of negative effect upon the embryo. Thus Nicolas Andry de Vois-regard, dean of the faculty of medicine in Paris in 1700, wrote: "If the pregnant woman's passionate desire [*envie*] for certain things she cannot obtain right away is sometimes capable of producing deformities in the child she is carrying, the sight of an object that causes her revulsion and horror is even more capable of doing so."[62] This works simply enough in cases where a woman is frightened by the sight of a deformed man and brings forth a deformed child, or by the sight of a knife and brings forth a child with a birthmark in the shape of a knife. But what if that repulsive "object" is her husband? Jewish and Christian texts do not seem to have devoted much attention to this possibility, but it is much discussed topic in ancient Indian texts, to which we now turn.

ANCIENT INDIA: VYASA AND VARUTHINI

The idea of parental imprinting is expressed in several Hindu myths, the most famous of which is closely associated with the Levirate, or *niyoga*—the duty of a dead man's brother to beget a son upon his brother's widow. The child of such a Levirate is regarded as the son of the dead man, whose widow presumably imagines him when she is in the arms of his brother. In the great Sanskrit epic the *Mahabharata*, composed over a period of several centuries before and after the turn of the common era, the sage Vyasa—a dirty old man who appears in the epic as a kind of walking semen bank—is called in to beget sons upon Ambika and Ambalika, the widows of his half-brother:

> When queen Ambika looked at Vyasa she saw his tawny matted hair and his blazing eyes and his red beard. "How ugly!" she thought; she was so frightened that she could not look at him, and in her terror she shut her eyes as tight as buds. Indeed, the sage was ugly, a skinny man of a most peculiar color. [He said,]

"Because of his mother's deficiency in the quality of sight, [the child] will be blind." And after a while, Ambika gave birth to a son who was blind, Dhritarashtra.

Ambalika sat on a splendid bed, deeply depressed, wondering, "Who is it who will come?" Then the great sage came to Ambalika in the same way. When she saw him, she too was so upset that she turned pale, and Vyasa said, "Since you with your lovely face turned pale when you saw how ugly I am, therefore this son of yours will be pale, and his name will be Pale (Pandu)."[63]

The text seems to take seriously—far more seriously than the texts that we have considered from other traditions—the woman's ability to imprint the child not with what she passively sees, or even actively fantasizes, but with what she *does* in reaction to what she sees; she closes her eyes (which we may see as simply another aspect of the emphasis on vision) or becomes pale. In fact, however, we may reformulate this as an instance of *paternal* impression: the child is born looking like *what the father saw* (and put into words in his curse)—a woman pale, or with her eyes closed.

The women reject Vyasa because he is old and ugly, but also because he is the wrong color (too dark? too light?), and this, plus Ambalika's temporary pallor, results in the birth of a child who is the wrong color, Pandu the Pale. Is this another echo of the racial aspect of color that haunts tales of maternal impression? Or is it an instance of the intersecting claims of maternal and paternal influence—the child who is pale both because his mother turned pale and because his father was the wrong color? Precisely such a conflation of influences is manifest in one of the few other ancient Indian stories of this genre, in a text composed several centuries after the *Mahabharata*. The woman in this story, like the mother of Dhritarashtra, closes her eyes. She does so not in conscious rejection of the man who forced himself upon her, but in imagination of the man she would have preferred—and thinks she is in bed with. This time, therefore, the child is born not blind but in the image of that other man:

Varuthini, a courtesan of the gods, fell in love with a Brahmin named Pravara and begged him to stay with her, but he rejected her and returned to his wife. Now, a demigod [*gandharva*] named

Kali was in love with Varuthini and had been rejected by her. He observed Varuthini now and reasoned, "She is in love with a human. If I take on his form, she will suspect nothing and will make love with me." He approached her and said, "You must not look at me during the time of our shared sexual enjoyment, but close your eyes and unite with me." She agreed, and when they made love, and her eyes were tightly closed, she thought, because of his hot semen, it was the form of [the Brahmin] suffused with the sacrificial fire. Then, after a while, she conceived an embryo, who came from the demigod's semen and from (her) thinking about the Brahmin's form. The demigod went away, still in the form of the Brahmin.[64]

The implication is that the demigod asks her to close her eyes because he fears he will reveal his true form when he makes love. But the text also implies that the child had the *true* form of the man that Varuthini thought she was making love with, just as, in the Zoharic midrash, Jacob's heir is the son of the woman he thinks he is in bed with when he begets his first son. The Sanskrit phrase, "from his semen and from her thinking," closely parallels the Hebrew phrase, "his semen followed his thought"— though with the essential difference that, here, his semen follows *her* thought.

A Telugu version of the story explains how the child could be affected in this way:

> Through that experience of deep delight
> the flame that was burning
> in Pravara's body
> became magically kindled
> in the *gandharva*'s form
> now held fast in *her*
> unwavering mind.
> Thus the child, a glowing fire,
> was conceived and grew to ripeness
> during nine months in her womb.[65]

Presumably the child then had not only Pravara's form but Pravara's flame, his essence, indeed his soul.

One medical text, composed sometime after the ninth century C.E., conflates external and internal influences: the child is said to resemble not only whatever creature the mother might be thinking of at the time of conception, but whatever she might look at during her fertile period, right after menstruation ends.[66] Thus, a woman who wanted to bear a white child was encouraged to furnish her bedroom with white things and, morning and evening, to look constantly at a big, white bull or a white horse of noble breeding—a new twist on the combination of animal husbandry, color coding, and visual imprinting.[67]

But there is another party that outranks the father, let alone the mother, in shaping the embryo: the embryo itself. For, according to the karma theory, the nature of a child is determined primarily not through the imaginations of his [*sic*, as usual] mother or father, but through his own imaginations in previous lives. Far more relevant than what the parent fantasizes in the sexual act is what the embryo thought about (or, in some cases, saw or even heard)[68] when he was dying. In terms of the theory of parental imprinting, the unborn embryo is his own parent, and imprints what he imagines or sees upon his future self.

Hindu embryology also assumes that the male embryo (one is not concerned about any other kind) is sentient within the womb, which further limits the agency of the woman in whose womb he sojourns for nine months. Medieval Hindu medical texts imagine the musings of the child inside the womb,[69] and several Tibetan and Indian Buddhist texts maintain that the state of mind of the unborn embryo (the transmigrating soul of someone who has recently died), hovering voyeuristically over the bed of the copulating parents-to-be, determines the nature (and, more specifically, the gender) of the baby who is to be born: if the embryo desires the woman and therefore hates the man, it becomes a male child; if the opposite, a female.[70] This proto-Freudian plot leaves little room for the intervention of the mother's mind in forming the embryo. We might call it embryonic imprinting.

CONCLUSION

There are clear historical links between the Jewish, Greek, and Christian texts we have considered, which explains some of their overlapping ideas. Some of these influences may have operated in India as well, but by and large the Indian examples function in contrast, making us aware, on the one hand, of the arbitrariness of much of what is shared in the other traditions and, on the other hand, of possibilities that the other traditions never seem to have considered.

Let us attempt to sort out a number of rather different conclusions that have been derived from this shared assumption of parental imprinting. First of all, the gender of the imprinter varies: occasionally a man will exert a force of paternal impression on a child, but overwhelmingly it is women who participate in this process. Second, there are differing views as to the moment at which imprinting takes place: the moment of conception, or any moment during pregnancy (or, in Hinduism, the period after menstruation or the moment of death). Premodern traditions favor the moment of conception. Third, the nature of the resulting child is seen in varying ways: some traditions regard it as monstrous, others as illegitimate but still human, still others as preferable to children born without such influences. But this difference must be correlated with a fourth factor: the causes of this veering from what is normal, primarily visual imprinting and mental imagination.

Visual imprinting, which reacts to material objects physically present at or before the time of conception, may take place accidentally or it may be brought about by the father on purpose, through the active manipulation of external stimuli and to his liking. For the Greeks, at least, the intentional use of images in a kind of pre-scientific planned-parenthood program overrode many anxieties about the failure of a child to resemble his parents. Here we must distinguish between the images that the husband offers to his wife in the hopes that she will desire a *child* cast in that image (though only Maimonides, among the sources cited here, took the logic one step further and suggested that the image should be of a child, not of a man), and the images that the wife may accidentally see that make her desire a *man*, and produce a child cast in his image. It comes down to the matter of who initiates the fantasy. Her fantasy is only acceptable if it is, in fact, *his* fantasy, his idea of what she should be seeing while he makes love to her; and it is certainly easier to regulate external vision than internal vision.

If imitation is the sincerest form of flattery, different traditions disagreed as to who should be flattered. Early Greek sources (and Paracelsus) imagined fathers who wanted their sons to be better than them, or at least more beautiful, and so resorted to the techniques of animal husbandry and the manipulation of works of art to influence the mind of the mother, encouraging their wives to improve the stock by "flattering" other men—or, preferably, gods. But Jews were in general aniconic as well as anti-eugenic. What appears to have been a mostly satisfying fiction in Hellenistic lore reappears in rabbinic literature as a disturbing possibility, requiring censure; the midrashim reject eugenics as bastardy. Most Christian sources similarly imagined fathers who wanted their sons to be just like them, to resemble no one else (except, sometimes, God). And Hindus, too, believed in the negative effects of impressions from images not intentionally presented by the father. They also believed that the embryo himself might exert mental influence, for ill or good, upon the physical form of his future self.

But certain tropes transcend cultural barriers. Male authors within all of the traditions we have touched upon feared the woman who imagined a man of her own choice, instead of just looking at a picture supplied by her husband. They insisted that the woman's gaze is passive; by seeing, she herself is imprinted, and this visual passivity overrides the active role she takes in handing this imprint on to her child, whereas a man would take the initiative from the start in actively stamping him with the paternal mark. By contrast, the imprinting that takes place entirely within the mother's mind, through her active imagination, excites the father's jealousy. Thomas Laqueur states the case very well: "Since normal conception is, in a sense, the male having an idea in the woman's body, then abnormal conception, the mola, is a conceit for her having an ill-gotten and inadequate idea of her own."[71] The eye was therefore an instrument of eugenics, passively and externally receiving the images produced by the father, while the mind (hidden inside the woman) was an instrument of adultery.

The ancient tales of black women giving birth to white children through the implicitly positive influence of works of art (positive, in the view of these texts, both because art is superior to nature and because white is superior to black) yield, in later cultures, to racist fantasies about white women cuckolding their husbands with black lovers. These racist images are easily conflated with the ancient mechanisms of animal husbandry, in which color is an essential factor, through the implicit equation of animals with people of other races.[72] The emphasis on color is a part of the general emphasis on physical (visual) resemblance, which usually made other factors such as behavior seem irrelevant in establishing paternity. The best way, therefore, to answer the question, "Whose child is it?" was to ask another question: "Whose face does it have?"

What seems most astonishing in all of this is the extent to which the seemingly most plausible explanation for the birth of a child who does not resemble his father—namely, the fact that some other man fathered him—is rejected by most of our sources (with the notable exception of the Jewish sources) in favor of fantasies about fantasies of a most extravagant nature. If your wife gives a birth to a child who looks just like your best friend, it needs no ghost come from the grave to explain the cause. Our premodern sources knew this, too. Yet though Heliodorus ("What had happened was so fantastic that no one would believe my explanation") and Voltaire ("This influence of the imagination is inexplicable") believed that the theory of maternal imprinting was unbelievable, they believed it nonetheless.

Even the woman's active imagination, far more threatening than her response to an image prepared by her husband, was not as threatening to her husband as a real man glimpsed in the street or market, even if he subsequently made love to the woman only in her fantasy and never in the flesh. The simpler idea (adultery) does seem to have occurred to some. Thus in 1726 Dr. James Blondel remarked of one of Malabranche's cases that the mother had lied and Malebranche was too naive to recognize this—an argument that, as Marie-Hélène Huet remarks archly, "had never been made by those disagreeing with Malebranche."[73] And Jean-Baptiste Demangeon wrote in 1807, "As for dissemblances, it is not unreasonable to believe that, when they are not prompted by adultery, they result from some disorder in the functioning of the organs of nutrition."[74] Yet this, too, is puzzling. Since these men believed that women did in fact lie (dissemble) and commit adultery all the time, why did they not invoke that belief in the context of embryonic dissemblance? The answer must lie in two rather different factors: their desperate need for some sort of assurance of paternity, and their genuine curiosity about what we would now call genetics, particularly about the problem of nonresemblance—a curiosity that transcended even the highly charged agenda of paternal insecurity.

The theory of maternal imprinting mutes the power of the real other man, the obvious villain of the piece; the complex mechanical causations of imprinting obscure the affective dimension of wives falling in love with other men. There may have been a real man, but she merely imagined him, through the adultery of thought. This is then further distanced by the suggestion that she merely imagined a man who does not in fact exist (such as a character from fiction or mythology), then that she imagined an artistic representation of a real or unreal man, and then that the husband takes charge and deliberately places a picture of another man in the bedroom for his wife to gaze at—as a model for the desired *child*—while he makes love to her. But the repressed knowledge of adultery is always there and bursts out in various paranoid forms. To men who fantasized that mental acts influenced the quality of their offspring, the very survival of the species depended upon the sexual fantasies of their women.

ENDNOTES

[1]Thomas Laqueur, *Making Sex: Body and Gender from the Greeks to Freud* (Cambridge, Mass.: Harvard University Press, 1990), 58.

[2]The Babylonian Talmud, compiled after the fourth century C.E., briefly mentions the same eugenic mechanism in a discussion of breeding the "red heifer without defect" required by Numbers 19:2 for the cultic preparations of the water sprinkled for impurity. To ensure the birth of an unblemished red calf, according to Rav Kahana, "They place a red cup in front of her [the cow] at the time when the male mounts her." (*Avodah Zerah* 24a. All references to the Babylonian Talmud are to the standard Vilna edition, and all translations from the Hebrew are by Gregory Spinner, unless otherwise noted.)

[3]Aristotle, *Problemata* 10.10, trans. W. S. Hett, Loeb Classical Library (1936). Aristotle discusses at some length the resemblance of children to their parents, in *De Partibus Animalium* 1.1, 640, trans. A. L. Peck, Loeb Classical

Library (1983), 63 and *De Generatione Animalium* 4.3, 767–769, trans. A. L. Peck, Loeb Classical Library (1983), 401–416.

[4]Aristotle, *De Partibus Animalium* 1.1, 640a:19–22 (p. 61) and *De Generatione Animalium* 4.3, 769a:15 (p. 415).

[5]Empedocles, cited by Aetius in *Doxographi Graeci* 5.12.2, ed. Herman Diels (Berlin: Walter de Gruyter, 1965), 432. See also *The Poem of Empedocles*, trans. Brad Inwood (Toronto: University of Toronto Press, 1992), 185.

[6]Soran, *Gynecology* 1, par. 39; in *Soranos d'Éphèse: Maladies des Femmes*, trans. Paul Burguière and Danielle Gouryevitch (Paris: Les Belles Lettres, 1988), 36; see also *Soranus' Gynecology*, trans. Oswei Temkin (Baltimore: Johns Hopkins, 1956), 37–38.

[7]Oppian, *Kynegetica*, 1.327–328, in *Oppian Colluthus Tryphiodorus*, trans. A. W. Mair, Loeb Classical Library (1928), 34–35.

[8]See Wendy Doniger O'Flaherty, *Women, Androgynes, and Other Mythical Beasts* (Chicago: University of Chicago Press, 1980), 149–212.

[9]Heliodorus, *Ethiopica*, book 4, chapter 8; "An Ethiopian Story," in B. P. Reardon, *Collected Ancient Greek Novels*, trans. J. R. Morgan (Berkeley and Los Angeles: University of California Press, 1989), 349–589, here 432–433.

[10]J. R. Morgan, in Reardon, *Collected Ancient Greek Novels*, 433, citing Achilles Tatius 3.7 and Philostratos *Imagines* 1.29.

[11]Fortunio Liceti, cited by Marie-Hélène Huet, *Monstrous Imagination* (Cambridge, Mass.: Harvard University Press, 1993), 23.

[12]Frank M. Snowden, Jr., in *Blacks in Antiquity: Ethiopians in Greco-Roman Experience* (Cambridge, Mass.: Harvard University Press, 1970), argues that the Greeks were innocent of racism, though he is regarded by some reviewers as naive. Evidence in his favor might include the positive Greek attitude to the black Athena and black Demeter as well as Poseidon's friendship with the Ethiopians among whom he feasts, oblivious to the needs of his Greek friends (Homer, *Odyssey*, 1.22, 5.282–288).

[13]*Soranos d'Éphèse*, Burguière and Gouryevitch, 83.

[14]Heliodorus, *Ethiopica*, Book 10, chapter 14, 568–569.

[15]Jerome, *Hebrew Questions*, on Genesis 33; *Saint Jerome's Hebrew Questions on Genesis*, trans. C. T. R. Hayward (Oxford: Clarendon Press, 1995), 67.

[16]This is puzzling unless we assume that the whole story, which is after all *about* a reactive back-formation ("What could have produced a child of the wrong color? Perhaps it was the sight of someone of the wrong color . . ."), is itself first expressed in a reactive back-formation ("Imagine if a black queen were as concerned about a white baby as we white people [like Quintilian] are concerned about a black baby . . .").

[17]Maimonides, *The Medical Aphorisms of Moses Maimonides*, trans. Fred Rosner (Haifa: The Maimonides Research Institute, 1989), 388, citing *De Theriaco ad Pisonem VI*.

[18]Bereshit Rabbah 73:10, ed. Theodor-Albeck, 854.

[19]This connection was made by Julius Preuss, *Biblisch-Talmudische Medezin* (1911), trans. Fred Rosner (Brooklyn: Hebrew Publishing, 1978), 392.

[20]The ordeal, called Sota, is described in Numbers 5:12–31 and in the Mishnaic tractate of that name.

[21]*Bemidbar Rabba* 9:34, in M. A. Mirkin, ed., *Midrash Rabba* (Tel Aviv: Yavneh, 1977), 213 ff.

[22]*Tanhuma*, Naso, 7, ed. Hanokh Zundel, 141.

[23]As, for example, in Genesis 41:35; Isaiah 3:6; and Psalms 8:7, 106:42.

[24]*Nedarim* 20b; cf. *Kallah* 50b and *Kallah Rabbati* 52a.

[25]*Baba Metzia* 84a; cf. also *Berakhot* 20a. The version in *Berakhot* omits the phrase "and learned," thus emphasizing the external form of the beautiful R. Yohanan.

[26]Daniel Boyarin, *Carnal Israel: Reading Sex in Talmudic Culture* (Berkeley: University of California, 1993), 216 ff.

[27]Seymour J. Cohen, *The Holy Letter: A Study in Jewish Sexual Morality* (Nashvale, N.J.: Aronson, 1993), 142–144.

[28]See Elliot Wolfson, "Woman—the Feminine as Other in Theosophic Kabbalah: Some Philosophical Observations on the Divine Androgyne," in Laurence Silberstein and Robert Cohn, eds., *The Other in Jewish Thought and History* New York: New York University Press, 1994), 201 n. 66.

[29]Isaiah Tishby, ed., *The Wisdom of the Zohar: An Anthology of Texts*, vol. 3, trans. David Goldstein, The Littman Library (Oxford: Oxford University Press, 1991), 1402–1403.

[30]*Kallah* 50b, *Kallah Rabbati* 52a.

[31]Tishby, *The Wisdom of the Zohar*, vol. 2, 646–649.

[32]At *Sota* 26b, the legal parameters of adultery are interpreted through the phrasing of Numbers 5:13.

[33]Tishby, *The Wisdom of the Zohar*, vol. 3, 1401.

[34]Huet, *Monstrous Imagination*, 81.

[35]Ibid., 95.

[36]Paracelsus, *Liber de Generatione Hominis*, in Franz Hartmann, ed., *The Life of Paracelsus, with the Substance of his Teachings* (San Diego: Wizards, 1986), 215–216.

[37]Paracelsus, *De Morbis Invisibilis*, in Hans Ranser, ed., *Schriften, Theophrasts von Hohenheim gennant Paracelsus* (Leipzig: Insel Verlag, 1921), 314–315. Translation by Wendy Doniger.

[38]Huet, *Monstrous Imagination*, 15.

[39]Ambrose Paré, *Toutes les oeuvres* (1585), book 24, 925–926, cited by Huet in *Monstrous Imagination*, 15.

[40]Huet, *Monstrous Imagination*, 71.

[41]Ibid., 73.

[42]Voltaire, "Imagination," in Denis Diderot and Jean d'Alembert, eds., *Encyclopedie, ou Dictionnaire raisonne des art, sciences et metiers*, vol. 8 (Paris: Briasson, 1751–65), 560–563, here 561.

[43]Montaigne, *The Complete Essays of Montaigne*, trans. Donald Frame (Stanford: Stanford University Press, 1965), 75. See also Huet, *Monstrous Imagination*, 13 and 19–20 and Thomas Laqueur, *Making Sex*, 129.

[44]Huet, *Monstrous Imagination*, 21.

[45]Midas Dekkers, *Dearest Pet: On Bestiality*, trans. Paul Vincent (London and New York: Verso, 1994), 83.

[46]See Joseph Garver, "Die Macht der Phantasie: Die 'heredity of influence' als literarische Thema," in *Saeculum* (1982) (3–4):287–311.

[47]James Joyce, *A Portrait of the Artist as a Young Man* (Harmondsworth: Penguin Books, 1992), 259.

[48]Johann Wolfgang von Goethe, *Elective Affinities*, trans. James Anthony Froude and R. Dillon Boylan (New York: Frederick Ungar, 1962), 224.

[49]E. T. A. Hoffmann, "The Doubles," in *The Tales of E. T. A. Hoffmann*, ed. and trans. Leonard J. Kent and Elizabeth C. Knight (Chicago: University of Chicago Press, 1969), 234–279, here 273.

[50]Ambrose Paré, *On Monsters and Marvels*, trans. Janis L. Pallister (Chicago: University of Chicago Press, 1982), 38–39. Céard said the story came from Hippocrates's *Opera*, section III, "De natura pueri," perhaps through Sylvius.

[51]Ambrose Paré, *Monsters*, 190 n. 47.

[52]Arthur Schnitzler, "Andreas Thameyers Letzer Brief," in *Gesammelte Werke, Erste Abteilung, Erzählende Schriften* (Berlin: S. Fischer Verlag, 1918), 220–228. Translation by Wendy Doniger.

[53]Havelock Ellis, "The Psychic State in Pregnancy," in *Studies in the Psychology of Sex*, vol. 5 (Philadelphia: F. A. Davis, 1906), 201–229, here 218.

[54]Montaigne, *The Complete Essays*, 578.

[55]Huet, *Monstrous Imagination*, 33.

[56]Ibid., 34 and 1.

[57]Ibid., 79–80.

[58]The same sort of triple-cross was at play in Jewish arguments that God could not only make the adulterer's child look like the adulterer (Vayikra Rabba 23:12) but could make him look like the husband, if the child was suspected of being fathered by someone else. Thus Rashi's midrash on Genesis 25:19 says that to counter rumors that Isaac had been fathered by Abhimelech, God intervened and miraculously changed Isaac's face into the image of Abraham's, so that the aged countenance of the father appeared on the face of the baby. (See also Baba Metzia 87a).

[59]Huet, *Monstrous Imagination*, 80.

[60]Ibid., 81.

[61]Eric Gerber, "Not-so-hot a Lover," Houston *Post*, 21 May 1986.

[62]Huet, *Monstrous Imagination*, 18.

[63]*Mahabharata* (Poona: Bhandarkar Oriental Research Institute, 1933–69) 1.99-100; Wendy Doniger O'Flaherty, *Textual Sources for the Study of Hinduism* (Chicago: University of Chicago Press, 1990), 46–51.

[64]*Markandeya Purana*, with commentary (Bombay: Venkatesvara Steam Press, 1890), chapters 58–61, esp. 59.30–31, 60.1–5.

[65]David Shulman, "First Man, Forest Mother: Telugu Humanism in the Age of Krsnadevaraya," in *Syllables of Sky: Studies in South Indian Civilization* (Delhi: Oxford University Press, 1995), 133–164, here 147, translating verse 4.3 and 5.20 of Allasani Peddana's *Manucaritramu*.

[66]Ram Karan Sharma and Vaidya Bhagwan Dash, eds., *Caraka Samhita*, text with commentary based on Cakrapani Datta's *Ayurveda Dipika*, Chowkhamba Sanskrit Series 44, vol. 2 (Varanasi: 1977), *Sharira Sthana, sutra* 25: "garbhopapattau tu manah sriyaa yam jantum vrajet tat saddrsham prasuute."

[67]*Caraka Samhita, Sharira Sthana*, 8.9.

[68]There is a medieval Sanskrit story about a virtuous sage who died in a hermitage; as he died, demons overran the hermitage, and people shouted, "Demons!" He heard this and was reborn as a demon. See Wendy Doniger O'Flaherty, ed., *Karma and Rebirth in Classical Indian Traditions* (Berkeley: University of California Press; Delhi, Motilal Banarsidass, 1980).

[69]*Markandeya Purana* 10.1–7, 11.1–21; O'Flaherty, *Textual Sources*, 97–98.

[70]J. P. McDermott, "Karma and Rebirth in Early Buddhism," in O'Flaherty, ed., *Karma and Rebirth in Classical Indian Traditions*, 165–192, here 171–172.

[71]Laqueur, *Making Sex*, 59.

[72]A case involving color was reported in the *Lancet* in 1890: A woman had been startled, when four months pregnant, by a black and white collie dog; she gave birth to a child whose "right thigh was encircled by a shining black mole, studded with white hairs." C. W. Chapman, *Lancet*, 18 October 1890; cited in Havelock Ellis, "The Psychic State in Pregnancy," 219. The mole is strongly reminiscent of the black birthmark in the form of a ring on Charikleia, in the tale told by Heliodorus.

[73]Huet, *Monstrous Imagination*, 66.

[74]Ibid., 78.

Among humanists it is commonly said it is useless to try to understand modern science because its language is so technical; scientific specialists do not now understand each other—and so, why should the humanists attempt what scientists themselves do not try to do? This argument is in one sense irresistible. In another sense it rests upon a remarkable assumption. The assumption is that the language of science is difficult, and the inference is that the language of humanism is always plain. I think, however, it is not unusual for humanists to remark with pride that Professor So-and-so is so learned his latest book can be understood only by ten or twelve in the nation, or in Europe, or in the world—the form of the legend varies. This may be mythical; but I can testify in sober fact from a good many years in the neighborhood of university presses that most books of humanistic scholarship cannot be understood even by the alumni formerly taught by the learned scholars. It makes a difference whose vocabulary is gored. . . . Every trade has its jargon, and I see no more reason to assume that humanists write as clearly as Bertrand Russell than I see reason to expect every biologist to write like William Kingdon Clifford.

—Howard Mumford Jones

"A Humanist Looks at Science,"
from *Dædalus* Winter 1958,
"Science and the Modern World View"

Edward O. Wilson

Consilience Among the Great Branches of Learning[1]

T HE CENTRAL THEME OF THE ENLIGHTENMENT, enhanced across three centuries by the natural sciences, is that all phenomena tangible to the human mind can be rationally explained by cause and effect. Thus humanity can—all on its own—know; and by knowing, understand; and by understanding, choose wisely.

The idea is amplified by what Gerald Holton has called the Ionian Enchantment, the conviction that all tangible phenomena share a common material base and are reducible to the same general laws of nature.[2] The roots of the Enchantment reach to the beginnings of Western science in the sixth century B.C., when Thales of Miletus, in Ionia, considered by Aristotle to be the founder of the physical sciences, proposed that all substances are composed ultimately of water. Although the hypothesis was spectacularly wrong, the ambition it expressed—to attain the broadest possible generalization in cause-and-effect explanations—was destined to become the driving force of Western science.

The success of the scientific revolution may make this perception now appear trivially obvious. Surely, it will seem to many, coherent cause-and-effect explanation is an inevitable consequence of logical thought. But to see otherwise it is only necessary to examine the history of Chinese science. From the first through the thirteenth centuries, as Europe passed from late antiquity through the Dark Ages, science in China flourished. It kept pace with Arab science, even though geographic isolation deprived Chinese scholars of the

Edward O. Wilson is Research Professor and Honorary Curator in Entomology at Harvard University.

ready-made base that Greek culture provided their Western counterparts. The Chinese made brilliant advances in subjects such as
descriptive astronomy, mathematics, and chemistry. But they never
acquired the habit of reductive analysis in search of general laws
that served Western science so well from the seventeenth century
on. They consequently failed to expand their conception of space
and time beyond what was attainable by direct observation with the
unaided senses. The reason, according to Joseph Needham, the
principal Western chronicler of the subject, was their emphasis on
the holistic properties and harmonious relationships of observable
entities, from stars to trees to grains of sand.[3] Unlike Western
scientists, they had no inclination to search for abstract codified law
in nature. Their reluctance was stimulated to some degree by the
historic rejection of the Legalists, who attempted to impose rigid,
quantified law during the transition from feudalism to bureaucracy
in the fourth century B.C. But of probably greater importance was
the fact that the Chinese steered away from the idea of a supreme
being who created and supervises a rational, law-governed universe. If there is such a ruler in charge, it makes sense—Western
sense at least—to read a divine plan and code of laws into physical
existence. If, on the other hand, no such ruler exists, it seems more
appropriate to search for separate rules and harmonious relations
among the diverse entities composing the material universe. In
summary, it can be said that Western scholars but not their Chinese
counterparts hit upon the more fortunate metaphysics among the
two most available to address the physical universe.

Western scientists also succeeded because they believed that
the abstract laws of the various disciplines in some manner
interlock. A useful term to capture this idea is *consilience*. The
expression is more serviceable than coherence or interconnectedness
because the rarity of its usage has preserved its original meaning,
whereas coherence and interconnectedness have acquired many
meanings scattered among a plethora of contexts. William
Whewell, in his 1840 synthesis *The Philosophy of the Inductive
Sciences*, introduced consilience as literally a "jumping together"
of facts and theory to form a common network of explanation
across the scientific disciplines. He said, "The Consilience of
Inductions takes place when an Induction, obtained from one
class of facts, coincides with an Induction, obtained from an-

other different class. This Consilience is a test of the truth of the Theory in which it occurs."

Consilience proved to be the light and way of the natural sciences. Physics, with its astonishing congruity to mathematics, came to undergird chemistry, which in turn proved foundational for biology. The successful union was not just a broad theoretical consistency, as articulated by Whewell, but an exact folding of principles pertaining to more complex and particular systems into the principles for simpler and more general systems. Organisms, it came to pass, can be reduced to molecules whose properties are entirely conformable to the laws of chemistry, and the elements to which the molecules are composed are in turn conformable to the laws of quantum physics.

To place the organization of modern science in clearer perspective, the disciplines can be tied to the position that their entities occupy in the scale of space and time, while noting that each class of entities represents a level of organization determined by the ensemble of other entities composing them and located lower on the space-time scale.

The consilient view of the natural world is illustrated by the use of the space-time scale to define the disciplines of biology:

Evolutionary space-time. Over many generations entire populations of organisms undergo evolution, which at the most elemental level is a change in the frequencies of the genes in the organisms that compose the populations. The foremost cause of evolution is natural selection, the differential survival and reproduction of the competing genes—or, put more precisely, the differential survival and reproduction of the organisms whose traits are determined by the genes. Natural selection occurs when populations interact with their environment. The subdiscipline broadly covering the phenomena in this segment of space-time is evolutionary biology.

Ecological space-time. Evolution by changes in gene frequency is coarse grained: It becomes apparent only when the history of an entire population is watched across generations. The process of natural selection driving it is finer grained, comprising particular events that affect the birth, reproduction, and death of individual organisms. These are events that can be observed only in a more constricted space and during shorter periods of time, usually the span of a season or less, than is the case for genetic evolution. They are addressed by the discipline of ecology. (Ecology is often put under the rubric of evolutionary biology, when that subject is broadly defined.)

Organismic space-time. Natural selection acts on the anatomy, physiology, and behavior of organisms whose programs of development are prescribed by genes. These properties usually occupy millimeters to meters in space and seconds to hours in time. The subdiscipline treating them is organismic biology.

Cellular space-time. The anatomy, physiology, and behavior of organisms are aggregated phenomena of cells and tissues. Covering micrometers to centimeters, and milliseconds to full generations, they are the province of cellular and developmental biology.

Biochemical space-time. The development and function of cells and tissues are themselves the aggregate products of highly organized systems of molecules. At this latter level, space ranges from nanometers to millimeters, and time usually from nanoseconds to minutes. The responsible discipline is molecular biology.

Two superordinate ideas unite and drive the biological sciences at each of these space-time segments. The first is that all living phenomena are ultimately obedient to the laws of physics and chemistry, with higher levels of organization arising by aggregate behavior at lower levels. The second is that all biological phenomena are products of evolution, and principally evolution by natural selection. The two ideas are expressions of consilience in the following way: Cells and thence organisms, being organized ensembles of molecules, are physicochemical entities, which were assembled not at random but by natural selection. Looked at this way, consilience in biology is the full sweep through the space-time scale, from near-instantaneous molecular process to the transgenerational shifts of gene frequency that compose evolution.

To many critics, especially in the social sciences and humanities, such an extreme expression of reductionism will seem fundamentally wrong-headed. Surely, they will say, we cannot explain something as complex as a brain or an ecosystem by molecular biology. To which most biologists are likely to respond, yes, we can, or we will be able to do so within a few years. The critics in turn call that impossible; such complex systems are distinguished by holistic, emergent properties not explicable by molecular biology, let alone atomic physics. The only fair response to this is yes, put that way, you are right.

Thus arises the paradox of emergence: Complex biological phenomena are reducible but cannot be predicted from a knowledge of molecular biology, at least not contemporary molecular biology. Each higher level of organization requires its own principles, including precisely definable entities, processes, spatial relationships, interactive forces, and sensitivity to external influences, which permit an accurate characterization and perhaps a stab at prediction from knowledge of its elements. Still, the principles, if sound, can be reduced from the top down and stepwise to those formulated at lower levels of organization. An ecosystem, to take the most complicated of all levels, can be broken into the species composing its biota. The species in turn can be analyzed according to the demography of the organisms composing them (population size and growth, birth and death schedules, age structure), along with their interactions with other kinds of organisms and with the physical environment. As part of this study, the organisms can be divided into organ systems, the organ systems into tissues and cells, and so on. The ecosystem, like other biological systems, is not truly hierarchical but heterarchical. It is constrained by the nature of its elements, and the behavior of the elements is determined at least in part by the sequences and proportions in which they are combined. By and large, however, the entities of each level can be reduced; and the principles used to describe the level, if apposite and correct, can be telescoped into those of lower levels and, especially, the next level down. That in essence is the process of reduction, or top-down consilience, which has been intellectually responsible for the enormous success of the natural sciences.

To proceed in the opposite direction, bottom-up, by synthesis—simple to more complex, general to more specific—is far more difficult. Physical scientists have succeeded splendidly at the task. They have interwoven principles of quantum theory, statistical mechanics, and reagent chemistry into stepwise syntheses from subatomic particles to atoms to chemical compounds. Advances in biology, if we measure their success by predictive power, have been much slower. Scanning the space-time scale along which biological complexity increases, we can see progress decelerate to a near stall at the level of protein synthesis. This is a critical juncture in the life sciences. About one hundred thousand kinds of protein molecules are found in the body of a vertebrate animal. Along with the nucleic acids that encode them, they are the essential materials of life. In particular, proteins form most of the basic structure of the body while running its machinery through catalysis of organic chemical reactions. Thanks to advances in technology, biochemists find it relatively easy to sequence the amino acids composing at least the smaller protein molecules, and to map the three-dimensional configuration in which these units are arrayed. It is another matter entirely, however, to predict how amino acids will fold together to create the configuration.[4] Three-dimensional form is all-important in the case of enzymes, which are the protein catalysts, because it determines which substrate molecules the enzyme molecule captures and which reaction it then catalyzes. When procedures are worked out to predict the exact shapes that arise from particular amino acid sequences, the result is likely to be a revolution in biology and medicine. It will permit the design of artificial enzymes and other proteins with desirable properties in biochemical reactions—perhaps superior to those occurring naturally. The difficulty is technical rather than conceptual: Prediction requires the integration of binding forces among all the amino acids simultaneously, an enormous computational problem; and in order to proceed that far it must also measure the forces with a precision beyond the capability of present-day biochemistry.

Even greater challenges are presented by the conceptual reconstruction of cells and tissues from a knowledge of the constituent molecules and chemical processes obtained through reductive analysis. In 1994 the editors of *Science* asked a hundred cellular and developmental biologists to identify the most important unsolved problems in their field of research. Their responses focused prominently on the mechanisms of synthesis.[5] In rank order, the problems most often cited were the following: 1) the molecular mechanisms of tissue and organ development; 2) the connection between development and genetic evolution; 3) the steps by which cells become committed to a particular fate during development; 4) the role of cell-to-cell signaling in tissue development; 5) the self-assembly of tissue patterns during development of the early embryo; and 6) the manner in which nerve cells establish their specific connections to create the nerve cord and brain. Although these problems are formidably difficult, the researchers reported that considerable progress has already been achieved and that the solution of several may be reached within a few years.

To summarize to this point, the consilience of material cause-and-effect explanations is approaching continuity throughout the natural sciences, binding them together across the full span of space and time. Of the two complementary processes of consilience, reduction and synthesis, the more successful has been reduction, because it is both conceptually and technically easier to master. Synthesis good enough to be quantitatively predictive has progressed much more slowly, but it is now inching its way within biology to the level of cell and tissue.

Yet despite the progress of the natural sciences in understanding the natural world, they have remained sequestered from the other great branches of learning. The social sciences and humanities are generally thought to be too grounded in ineffable phenomena of mind and culture, too complex and holistic, and too dependent on historical circumstance to be consilient with the natural sciences.

That venerable perception, I believe, is about to change. The reason is that the natural sciences, doubling in information content every two decades or less, have now expanded to touch the material processes that generate mental and cultural phenomena. Two disciplines—the brain sciences and evolutionary biology—are now filling the ancient gap between dual epistemologies to serve as bridges between the great branches of learning.

The brain sciences are a conglomerate of research activities by neuroscientists, cognitive psychologists, and philosophers ("neurophilosophers") bound together by their conviction that the mind is the brain at work and, as such, can be understood entirely as a biological phenomenon. For their part, evolutionary biologists address the origin of the mental process, which is also considered a biological process. In particular, they focus on the instinct-like emotional responses and learning biases that affect individual development and the evolution of culture.

The key and largely unsolved problem of the brain sciences is the neuron circuitry and neurotransmitter fluxes composing conscious thought. The most important entrée to the problem is brain imaging, the monitoring of brain activity by the direct mapping of its metabolic patterns. The current method of choice in brain imaging is positron emission tomography (PET) scanning, which measures activity in different parts of the brain by the amount of their blood flow—hence the oxygen and energy being delivered to them. The patient is first injected with a small amount of rapidly decaying isotope of oxygen or another harmless radioactive material that emits elementary particles called positrons. The positrons interact with electrons in tissue reached by the isotope, resulting in radiation that can be picked up by a camera. As the patient experiences a sensation, or reflects upon a subject, or feels an emotion, blood flow increases within a tenth of a second in the activated part of his brain, and the corresponding change is detected by the scanner.

An alternative method of brain imaging is functional magnetic resonance imaging (fMRI). Its precursor recording method is static magnetic resonance imaging (MRI), which is based on the response of molecules in body tissues to radio waves after the molecules have been forced into a certain orientation by a powerful magnet. The magnitude of the response rises according to the water content of the tissues, which in turn increases while blood (half of which is water) flows into the active areas. Researchers convert MRI into fMRI, which enables them to use it to monitor brain activity, by recording multiple images through time. The images are then viewed in rapid succession to create moving images in the manner of conventional cinematography. The fMRI method is more efficient in this respect than PET scanning, having been improved to record hundreds of images per minute.

As in all biological research, the overall evolution of brain scanning is toward ever deeper, finer, and faster probes of activity. Other methods directed toward these goals, based on different physical phenomena from those employed in PET and fMRI, have recently opened a new chapter in imaging technology. One method, still limited currently to experimental use in animals, is the application of voltage-sensitive dyes to the surface of the living brain. The electrical conduction of the nerve fibers literally light up the dyes in patterns that can be tracked by photodiode cameras. Images have been recorded in excess of a thousand per second, allowing more nearly continuous monitoring than PET and fMRI scanning.

As the twenty-first century opens, we can expect to witness the invention of even more sophisticated methods of brain imaging, as well as refinement of those already in use. With luck, scientists will eventually reach their ultimate goal of monitoring the activity of intact brains continuously and at the level of individual nerve fibers. In short, the mind as brain-at-work can be made visible.

Brain imaging and experimental brain surgery, together with analyses of localized brain trauma and endocrine and neurotransmitter mediation, have permitted a breakout from age-old subjective conceptions of mental activity. Researchers now speak confidently of a coming solution to the brain-mind problem.

Some students of the subject, however (including a few of the brain scientists themselves), consider that forecast overly optimistic. In their view, technical progress has been largely correlative and has contributed little to a deeper understanding of the conscious mind. They consider it the equivalent of mapping the communicative networks of a city, correlating its activity with ongoing social events, and then declaring the material basis of culture solved. Even if brain activity is mapped completely, they ask, where does that leave consciousness, and especially subjective experience? How to express joy in a summer rainbow with neurobiology? Perhaps these phenomena rise from undiscovered physicochemical phenomena or exist at a level of organization still beyond our comprehension. Or maybe, as a cosmic principle, the conscious mind is just too complicated and subtle ever to understand itself.

This view of the mind as *mysterium tremendum* is, in the opinion of most brain researchers, unjustifiably defeatist. It is the residue of mind-body dualism, the impulse to posit a master integrator—whether corporeal or ethereal—located somewhere in the brain and charged with integrating information from the neural circuits and making decisions. The perception weighs too lightly the alternative and more parsimonious hypothesis: That activity of the neural circuits *is* the mind, and as a consequence nothing more of fundamental aspect is needed to account for mental phenomena at the highest levels. In this view, the hundred million or so neurons, each with an average of thousands of connections to other neurons, are enough to symbolize the thick stream of finely graded information and emotional coloring we introspectively recognize as composing the conscious mind.

To envision the immense amount of information that can be encoded, consider the following hypothetical example supplied by neurobiologists. Suppose that the chemoreceptive brain were programmed to sort and retrieve information by vector coding. Suppose further that combined activities of nerve cells imposing the codes classify individual tastes into combinations of sweetness, saltiness, and sourness. The brain need only distinguish 10 degrees in each of these taste dimensions to discriminate $10 \times 10 \times 10$ or 1,000 substances.

A large part, if not the totality, of mental activity comprises scenarios built with such symbolic information. The scenarios are usually reconstructions of the here and now, during which the brain is flooded with fresh sensory information. Many others recreate the past as it is summoned from long-term memory banks. Still others construct alternative possible futures, or pure fantasy.

According to the parsimonious theory of mind, emotions are the modifications of neural activity that animate and focus the scenarios. An act of decision is the prevalence of certain future scenarios over others; those that prevail are most likely to be the ones most conformable to instinct and reinforcement from prior experience. What we think of as meaning is the linkage among neural networks. Learning is the spreading activation that enlarges imagery and engages emotion. The self (to continue the parsimonious theory) is the key dramatic character of the scenarios. It must exist, because the brain is located within the body, and the body is the constant intense focus of real-time sensory experience and decision making.[6]

The primary environment in which the mind develops is culture. This highest level of human activity was defined in 1952 by Alfred Kroeber and Clyde Kluckhohn, out of a review of 164 prior definitions, as follows: "Culture is a product; is historical; includes ideas, patterns, and values; is relative; is learned; is based upon symbols; and is an abstraction from behavior and the products of behavior."[7] It comprises the life of a society, the totality of its religion, myths, art, technology, sports, and all the other systematic knowledge transmitted across generations.

Throughout this century scholars in all the branches of learning have treated culture as an entity apart, comprehensible only on its own terms and not those of the natural sciences. By this conception culture stands apart even if the mind has a reducible, material basis; it must do so first because the fine details of the cultures of individual societies are historically determined, and second because cultures comprise phenomena too complicated, too flickering through time, and too subtle to be subject to natural scientific analysis.

A fixed belief in the independent nature of culture has contributed to the isolation of the social sciences and humanities from the natural sciences throughout modern history. It is the basis of the discontinuity famously cited by C. P. Snow in 1959 as separating the scientific culture from the literary culture. Now there is reason to believe that the difference is not a true epistemological discontinuity, not a divide between two kinds of reality, but something far less forbidding and yet much more interesting. The boundary between the two cultures is instead a vast, unexplored terrain of phenomena awaiting entry from both sides.

The terrain is the interaction between genetic evolution and cultural evolution. We know that culture is learned. At the same time, evidence is mounting that learning is genetically biased; it is becoming increasingly accepted that culture is influenced by human nature. But what exactly is human nature? It is not the genes, which prescribe it, or the cultural universals, which are its most obvious products. It is the epigenetic rules, the hereditary biases that guide the development of individual behavior. There are several examples of epigenetic rules that can be cited in this early stage of investigation.

The facial expressions denoting the elementary emotions of fear, loathing, anger, surprise, and happiness are human universals and evidently inherited. They are adjusted by cultural evolution within individual societies to project particular nuances of meaning. The smile, one of the basic elements of emotive communication, appears at two to four months in infants everywhere, virtually independent of environment. It occurs on schedule in deaf-blind infants and even in thalidomide-crippled children who cannot touch their own faces.[8] The tendency to fear snakes is another human universal. It is furthermore widespread, if not universal, in all other Old World primate species. Snakes are among the few stimuli that easily evoke true phobias in people—the deep and intractable visceral reactions acquired with only one or two frightening experiences. They share their power with heights, closed spaces, running water, spiders, and other ancient perils of humanity; a similar degree of sensitivity does not exist for knives, guns, electric sockets, automobiles, and other modern sources of risk. The cultural consequences of the response to snakes, combining fear and intense curiosity, are manifold. Snakes are among the animals most commonly experienced in dreams, even among urbanites who have never seen one in life. They play prominent mythic roles in cultures around the world, taking new forms variously as demons, dragons, seducers, magical healers, and gods.[9]

Automatic incest avoidance is universal in primate species studied to date, including *Homo sapiens*. The generally accepted adaptive explanation is the heightened risk that inbreeding poses of producing defective offspring, and that evolutionary inference is well supported by the evidence. The closer the genetic relationship of parents, the more likely they will bring together matching recessive genes that are deleterious in a double dose. Children of full siblings and of fathers and daughters, for example, have twice the early mortality rate of outbred children. Among those that survive, ten times more suffer genetic defects such as heart deformities, deaf-mutism, mental retardation, and dwarfism. The epigenetic rules, or hereditary developmental biases that prevent incest, are two-layered in apes, monkeys, and other non-human primates. First, all species so far studied for the trait (nineteen worldwide) practice the equivalent of human exogamy: Young individuals leave the parent group and join another before they attain full maturity. Second, all species examined for the possible existence of the Westermarck effect also display that phenomenon. This means that individuals are sexually desensitized to individuals with whom they have been closely associated while very young, normally their parents and siblings. The critical period for the effect in human beings is the first thirty months of life. Out of the Westermarck effect have apparently risen incest taboos with all their supporting arsenal of legends and myths. The effect is enhanced in some but not all societies by a third barrier: the direct observation and correct rational understanding of the ill effects of incest.[10]

Similar examples of epigenetic rules have multiplied in the literature of biology and the behavioral sciences during the past several decades. They have been found in virtually all categories of human behavior, including sexual and parental bonding, the acquisition of language, and even the cardinal role of trust during contract formation. They leave little doubt that a true hereditary human nature exists, and that it includes social behaviors held in common with nonhuman primate species and others that are diagnostically human.

Such is the interdisciplinary subject awaiting study by all the great branches of learning, and I can think of no more important intellectual undertaking. The relation between biological evolution and cultural evolution is, in my opinion, both the central problem of the social sciences and humanities and one of the great remaining problems of the natural sciences.

The process by which genetic evolution and cultural evolution appear to be linked is usually called gene-culture coevolution. The theory of gene-culture coevolution incorporates the two levels of approach I cited earlier as the core of modern biology.[11] Put as briefly as possible, they are that living processes are physicochemical and also self-assembled by natural selection. The first level is composed of proximate explanations, which describe the structures and processes by which an organism responds. The question of interest in any proximate explanation is, How does the phenomenon occur? The second level is composed of ultimate, or evolutionary, explanations, which account for the origin of the structures and processes, usually by the adaptive advantage they confer on organisms. The question of interest at this level is, *Why* does the phenomenon occur? In the case of hereditarily based incest avoidance, the proximate causes are emigration and the Westermarck effect. The ultimate cause is the deleterious effects of inbreeding, which by natural selection has driven the species toward emigration and the Westermarck effect.

The theory of gene-culture coevolution is still spotty and largely untested. Nevertheless, I believe that most researchers on the subject would agree with the following outline of the present form of the theory: People survive and leave offspring to the degree that they learn and adapt to the culture of their society, and the societies themselves flourish or decline in proportion to the effectiveness of their adaptation to their environment and surrounding societies. For hundreds of millennia certain aptitudes and cultural norms have arisen that are consistently adaptive in this Darwinian sense. They include language facility, cooperativeness within the group, exogamy and incest avoidance, rites of passage, territoriality, male polygyny, and parent-offspring bonding. Hereditary epigenetic rules have evolved that pull individual preference, and hence cultural evolution, toward these norms. They comprise the elements of what we subjectively call human nature. The genes prescribing them also increase in frequency as a result of the same process. Spreading through the population, maintained by the edge they give most of the time in survival and reproduction, they have secured the stability of human nature across societies and generations.

To conclude my synopsis of the theory, cultural evolution is much faster than genetic evolution. One result is nongenetic cultural diversity, which scatters particular cultural variants around each central, genetic trend to a degree determined by the strength of the epigenetic rules affecting them. The products of cultural evolution, multiplying rapidly through the population, can improve the fitness of individuals and societies, or they can reduce them. But only if the advantage or disadvantage is sustained for many generations—population genetics theory would suggest at least ten—can the epigenetic rules and the genes prescribing them be replaced. That is why human nature today remains Paleolithic even in the midst of accelerating technological advance. Thus corporate CEOs impelled by stone-age emotions work international deals with cellular telephones at thirty thousand feet.

If it is granted that the human condition is subject to consilient explanation from genes to mind to culture, even as a working hypothesis, the consequences to follow will be considerable. The first is support across the great branches of learning for what can appropriately be called "gap analysis" as a research strategy.[12] Already a mainstay of the natural sciences, gap analysis is the systematic attempt to identify domains of phenomena in which important discoveries are most likely to be made. Its most productive method is reduction, the search for novel phenomena, or at least the search for novel explanations of phenomena already known, by examination of the next level of organization down. Successful reduction confirms the existence of elements in the lower level that interact to create the higher level. In this manner, molecular biology was created de novo from the basic chemistry of macromolecules, and the study of cells and tissues was revolutionized by molecular biology.

The social sciences, I believe, will advance more rapidly if they adopt a consilient worldview and the gap analysis suggested by it leading to reductionist analysis. They have failed to give this approach a try, except in a few sectors such as biological anthropology, largely because of their aversion to biology. The reasons for the aversion are complex, stemming partly from the effort of the social science disciplines—anthropology, economics, political science, and sociology—to maintain intellectual independence, partly from the daunting complexity of the subject, and partly from fear of the misuse of biology to support racist ideology.

Still, biology is the logical foundational discipline of the social sciences. I mean by this assessment biology as broadly defined, including much of contemporary psychology, especially cognitive psychology, which is in the process of being subsumed by neurobiology and the brain sciences. A great majority of social scientists, including the most influential theoreticians in economics, build their models as if this information does not exist. Their conceptions of human behavior come either from folk psychology—intuitive notions that seem right but are often factually wrong—or from notions of the mind as an optimizing device for rational choice. They ignore contrary signs from genetics, neurobiology, cognitive psychology, and the many quirky properties of human nature. For them history began a few thousand years ago with the rise of complex societies, overlooking the fact that it began hundreds of thousands of years ago with the evolutionary origins of human nature in hunter-gatherer bands.

In summary, it is hard to imagine how the social sciences can unite and achieve general, predictive theory without taking a reductionist approach to the phenomena of human nature, both their proximate causes in the machinery of the brain and their ultimate causes in deep, evolutionary history.

The theory and criticism of the arts can also benefit in the same fashion. Let me cite several examples already in hand. We now know, from neurobiology and the brain sciences, how the brain breaks down and classifies the continuously varying wavelength of visible light into four basic colors, namely, blue, green, yellow, and red. The process has been tracked in segments from the base sequences in the DNA that prescribe the cone pigments of the photosensitive retinal cells to the nerve-cell sequences that lead from the retina to the primary visual cortex at the extreme rear of the brain. From anthropological and linguistic studies we know that people in societies around the world fix their color terms toward the centers of the primary colors in the spectrum and away from the intermediate and hence ambiguous wavelengths. Finally, we know that as societies increase their color vocabularies, in the course of cultural evolution, they tend to employ up to eleven basic terms, usually accumulating them in the following sequence: Languages with only two basic color terms use them to distinguish black and white; languages with only three terms identify black, white, and red; languages with only four terms have words for black, white, red, and either green or yellow; languages with only five terms have words for black, white, red, green, *and* yellow; and so on until all eleven terms are included, as exemplified in the English language. The sequence cannot be due to chance alone. If the terms were combined at random, there would be 2,036 possible combinations. But for the most part they are drawn from only 22. Surely this is the kind of information needed to produce a coherent theory of aesthetics in the visual arts.[13]

In another domain relevant to visual aesthetics, neurobiological measurements have shown that the brain is most aroused by abstract designs in which there is about 20 percent repetition of elements. That is the amount of redundancy found in a simple maze, two turns of a logarithmic spiral, or an asymmetrical cross. It seems hardly a coincidence that roughly the same property is shared by a great deal of the art in friezes, grillwork, colophons, and flag designs. Or that it crops up again in the glyphs of ancient Egypt and Mesoamerica as well as the pictographs of Japanese, Chinese, Thai, Bengali, and other Asian languages. The response appears to be innate: Newborn infants gaze longest at figures with about the same amount of redundancy.[14]

In yet another topic of aesthetics, ideal female facial beauty as judged in at least two cultures, European and Japanese, has recently been found to follow some surprising principles. Using blended and artificially altered photographs, psychologists have discovered that the most admired facial features are near the anatomical average of the population but with heightened cheekbones, reduced chin size, enlarged eyes, and shortened distance between the nose and chin.[15] The cause of this effect, if upheld as inborn by further cross-cultural and developmental studies, is unknown. It could represent an innate recognition of the signs of youthfulness and hence greater reproductive potential.

The creative arts themselves, in literature, the visual arts, drama, music, and dance, may not be affected significantly by such knowledge from the natural sciences. The purpose of the arts is to transmit personal experience and emotion directly from mind to mind while avoiding explanation of the logic behind the creative work; thus, *ars est celare artem*, it is art to conceal art. But theory and criticism of the arts, which does attempt this mode of explanation, cannot help but be strengthened by the new information. If the greatest art is indeed that which touches all humanity, as commonly said, it follows that consilient cause-and-effect accounts of human nature will become increasingly foundational to sound theory and criticism.

ENDNOTES

[1]This essay presents in much abbreviated form some of the arguments in my book-length exposition of the same general subject, *Consilience: The Unity of Knowledge* (New York: Knopf, 1998).

[2]The Ionian enchantment is discussed by Gerald Holton in *Einstein, History, and Other Passions* (Woodbury, N.Y.: American Institute of Physics Press, 1995).

[3]*The Shorter Science and Civilisation in China: An Abridgment of Joseph Needham's Original Text*, Vol. I, prepared by Colin A. Ronan (New York: Cambridge University Press, 1978).

[4]In characterizing the prediction of three-dimensional protein structure, I benefited greatly from an unpublished paper presented by S. J. Singer at the American Academy of Arts and Sciences in December 1993; he has also kindly reviewed my account.

[5]On the opinions of cell and developmental biologists concerning the frontiers of their field, see "Looking to Development's Future," by Marcia Barinaga, *Science* 266 (1994): 561–564.

[6]Among the many recent works I have used to interpret the consensus of students of the mind-body problem are Patricia S. Churchland, *Neurophilosophy: Toward a Unified Science of the Mind-Brain* (Cambridge, Mass.: MIT Press, 1986); Paul M. Churchland, *The Engine of Reason, the Seat of the Soul* (Cambridge, Mass.: MIT Press, 1995); Antonio R. Damasio, *Descartes' Error: Emotion, Reason, and the Human Brain* (New York: G. P. Putnam, 1994); Daniel C. Dennett, *Consciousness Explained* (Boston: Little, Brown, 1991); J. Allan Hobson, *The Chemistry of Conscious States: How the Brain Changes Its Mind* (Boston: Little, Brown, 1994); and Stephen M. Kosslyn and Oliver Koenig, *Wet Mind: The New Cognitive Neuroscience* (New York: Free Press, 1992).

[7]Alfred Kroeber and Clyde K. M. Kluckhohn, "Culture: A Critical Review of Concepts and Definitions," *Papers of the Peabody Museum of American Archaeology and Ethnology, Harvard University*, vol. 47 (1952), no. 12, 643–644.

[8]On basic facial expressions: the literature, including smiling, is reviewed by Charles J. Lumsden and Edward O. Wilson in *Genes, Mind, and Culture* (Cambridge, Mass.: Harvard University Press, 1981) and by the pioneer behavioral biologist Irenäus Eibl-Eibesfeldt in *Human Ethology* (Hawthorne, N.Y.: Aldine de Gruyter, 1989).

[9]On the fear of snakes and the origin of the serpent myth: Balaji Mundkur, *The Cult of the Serpent: An Interdisciplinary Survey of Its Manifestations and Origins* (Albany, N.Y.: State University of New York Press, 1983) and Edward O. Wilson, *Biophilia* (Cambridge, Mass.: Harvard University Press, 1984).

[10]On incest and its avoidance in human beings and other primates: Arthur P. Wolf, *Sexual Attraction and Childhood Association: A Chinese Brief for Ed-*

ward Westermarck (Stanford, Calif.: Stanford University Press, 1995) and William H. Durham, *Coevolution: Genes, Culture, and Human Diversity* (Stanford, Calif.: Stanford University Press, 1991).

[11]The expression gene-culture coevolution and a first general theory pertaining to it, in the sense of combining models from genetics, psychology, and anthropology, were provided by Lumsden and Wilson in *Genes, Mind, and Culture.* A review and update of the subject are given in my more general book *Consilience.*

[12]"Gap analysis" is a term I have borrowed from conservation biology. It means the method of mapping known ranges of threatened plant and animal species and using the information to select the best sites to set aside as reserves. See J. Michael Scott and Blair Csuti, "Gap Analysis for Biodiversity Surveys and Maintenance," in Marjorie L. Reaka-Kudla et al., eds., *Biodiversity II: Understanding and Protecting Our Biological Resources* (Washington, D.C.: Joseph Henry Press, 1997), 321–340.

[13]A full account of the biological and cultural origins of color perception and vocabulary is given by multiple authors in Trevor Lamb and Janine Bourriau, eds., *Colour: Art & Science* (New York: Cambridge University Press, 1995).

[14]On the optimum amount of redundancy in design: see the review by Charles J. Lumsden and Edward O. Wilson, *Promethean Fire* (Cambridge, Mass.: Harvard University Press, 1983).

[15]On female facial beauty: "Facial Shape and Judgements of Female Attractiveness," by D. I. Perrett et al., *Nature* 368 (1994): 239–242. Other aspects of ideal physical characteristics are discussed by David M. Buss in *The Evolution of Desire* (New York: BasicBooks, 1994).

Let us then not take science for what some of its philosophers would like it to be. Completely "intersubjective" statements from which all metaphysics is banished, from which any implication or discussion of being is removed, could occur only at the point where there are no subjects left to share them. So long as it is alive and not sterile, science will remain a *speculum entis*, it will present what metaphysics did, a symbolic structure which is an essential metaphor of being, but is not the only one.

—Giorgio de Santillana

"The Seventeenth-Century Legacy:
Our Mirror of Being,"
from *Dædalus* Winter 1958,
"Science and the Modern World View"

Steven Weinberg

Physics and History

I AM ONE OF THE FEW CONTRIBUTORS to this issue of *Dædalus* who is not in any sense a historian. I work and live in the country of physics, but history is the place that I love to visit as a tourist. Here I wish to consider, from the perspective of a physicist, the uses that history has for physics, and the dangers both pose to each other.

I should begin by observing that one of the best uses of the history of physics is to help us teach physics to nonphysicists. Although many of them are very nice people, nonphysicists are rather odd. Physicists get tremendous pleasure out of being able to calculate all sorts of things, everything from the shape of a cable in a suspension bridge to the flight of a projectile or the energy of the hydrogen atom. Nonphysicists, for some reason, do not appear to experience a comparable thrill in considering such matters. This is sad but true. It poses a problem, because if one intends to teach nonphysicists the machinery by which these calculations are done, one is simply not going to get a very receptive class. History offers a way around this pedagogical problem: everyone loves a story. For example, a professor can tell the story (as I did in a book and in courses at Harvard and Texas) of the discovery of the subatomic particles—the electron, the proton, and all the others.[1] In the course of learning this history, the student—in order to understand what was going on in the laboratories of J. J. Thomson, Ernest Rutherford, and our other heroes—has to learn something about how particles move

Steven Weinberg is Josey Regental Professor of Science at the University of Texas at Austin.

under the influence of various forces, about energy and momentum, and about electric and magnetic fields. Thus, in order to understand the stories, they need to learn some of the physics we think they should know. It was Gerald Holton's 1952 book *Introduction to Concepts and Theories of Physical Science* that first utilized precisely this method of teaching physics; Holton told the story of the development of modern physics, all the while using it as a vehicle for teaching physics. Unfortunately, despite his efforts and those of many who came after him, the problem of teaching physics to nonphysicists remains unsolved. It is still one of the great problems facing education—how to communicate "hard sciences" to an unwilling public. In many colleges throughout the country the effort has been given up completely. Visiting small liberal-arts colleges, one often finds that the only course offered in physics at all is the usual course for pre-medical students. Many undergraduates will thus never get the chance to encounter a book like Holton's.

History plays a special role for elementary particle physicists like myself. In a sense, our perception of history resembles that of Western religions, Christianity and Judaism, as compared to the historical view of other branches of science, which are more like that of Eastern religious traditions. Christianity and Judaism teach that history is moving toward a climax—the day of judgment; similarly, many elementary particle physicists think that our work in finding deeper explanations of the nature of the universe will come to an end in a final theory toward which we are working. An opposing perception of history is held by those faiths that believe that history will go on forever, that we are bound to the wheel of endless reincarnation; likewise, particle physicists' vision of history is quite different from that of most of the sciences. Other scientists look forward to an endless future of finding interesting problems—understanding consciousness, or turbulence, or high temperature superconductivity—that will go on forever. In elementary particle physics our aim is to put ourselves out of business. This gives a historical dimension to our choice of the sort of work on which to concentrate. We tend to seek out problems that will further this historic goal—not just work that is interesting, useful, or that influences other fields, but work that is historically progressive, that moves us toward the goal of a final theory.

In this quest for a final theory, problems get bypassed. Things that once were at the frontier, as nuclear physics was in the 1930s, no longer are. This has happened recently to the theory of strong interactions. We now understand the strong forces that hold the quarks together inside the nuclear particles in terms of a quantum field theory called quantum chromodynamics. When I say that we understand these forces, I do not mean that we can do every calculation we might wish to do; we are still unable to solve some of the classic problems of strong interaction physics, such as calculating the mass of the proton (the nucleus of the hydrogen atom). A silly letter in *Physics Today* recently asked why we bother to talk about speculative fundamental theories like string theory when the longstanding problem of calculating the mass of the proton remains to be solved. Such criticism misses the point of research focused on a historical goal. We have solved enough problems using quantum chromodynamics to know that the theory is right; it is not necessary to mop up all the islands of unsolved problems in order to make progress toward a final theory. Our situation is a little like that of the United States Navy in World War II: bypassing Japanese strong points like Truk or Rabaul, the Navy instead moved on to take Saipan, which was closer to its goal of the Japanese home islands. We too must learn that we can bypass some problems. This is not to say that these problems are not worth working on; in fact, some of my own recent work has been in the application of quantum chromodynamics to nuclear physics. Nuclear forces present a classic problem—one on which I was eager to work. But I am not under the illusion that this work is part of the historical progress toward a final theory. Nuclear forces present a problem that remains interesting, but not as part of the historical mission of fundamental physics.

If history has its value, it has its dangers as well. The danger in history is that in contemplating the great work of the past, the great heroic ideas—relativity, quantum mechanics, and so on—we develop such reverence for them that we become unable to reassess their place in what we envision as a final physical theory. An example of this is general relativity. As developed by

Einstein in 1915, general relativity appears almost logically inevitable. There was a fundamental principle, Einstein's principle of the equivalence of gravitation and inertia, which says that there is no difference between gravity and the effects of inertia (like centrifugal force). The principle of equivalence can be reformulated as the principle that gravity is just an effect of the curvature of space and time—a beautiful principle from which Einstein's theory of gravitation follows almost uniquely. But there is an "almost" here. To arrive at the equations of general relativity, Einstein in 1915 had to make an additional assumption; he assumed that the equations of general relativity would be of a particular type, known as second-order partial differential equations. This is not the place to explain precisely what a second-order partial differential equation is—roughly speaking, it is an equation in which appear not only things like gravitational fields, and the rates at which these things change with time and position, but also second-order rates, the rates at which the rates change. It does not include higher order rates, for instance, third-order rates—the rates at which the rates that are changing are changing. This may seem like a technicality, and it is certainly not a grand principle like the principle of equivalence. It is just a limit on the sorts of equations that will be allowed in the theory. So why did Einstein make this assumption—this very technical assumption, with no philosophical underpinnings? For one thing, people were used to such equations at the time: the equations of Maxwell that govern electromagnetic fields and the wave equations that govern the propagation of sound are all second-order differential equations. For a physicist in 1915, therefore, it was a natural assumption. If a theorist does not know what else to do, it is a good tactic to assume the simplest possibility; this is more likely to produce a theory that one could actually solve, providing at least the chance to decide whether or not it agrees with experiment. In Einstein's case, the tactic worked.

But this kind of pragmatic success does not in itself provide a rationale that would satisfy, of all people, Einstein. Einstein's goal was never simply to find theories that fit the data. Remember, it was Einstein who said that the purpose of the kind of physics he did was "not only to know how nature is and how her transactions are carried through, but also to reach as far as possible the utopian and seemingly arrogant aim of knowing why nature is thus and not otherwise. . . ." He certainly was not achieving that goal when he arbitrarily assumed that the equations for general relativity were second-order differential equations. He could have made them fourth-order differential equations, but he did not.

Our perspective on this today, which has been developing gradually over the last fifteen or twenty years, is different from that of Einstein. Many of us now regard general relativity as nothing but an effective field theory—that is to say, a field theory that provides an approximation to a more fundamental theory, an approximation valid in the limit of large distances, probably including any distances that are larger than the scale of an atomic nucleus. Indeed, if one supposes that there really are terms in the Einstein equations that involve rates of fourth or higher order, such terms would still play no significant role at sufficiently large distances. This is why Einstein's tactic worked. There is a rational reason for assuming the equations are second-order differential equations, which is that any terms in the equations involving higher order rates would not make much of a difference in any astronomical observations. As far as I know, however, this was not Einstein's rationale.

This may seem rather a minor a point to raise here, but in fact the most interesting work today in the study of gravitation is precisely in contexts in which the presence of higher-order rates in the field equations would make a big difference. The most important problem in the quantum theory of gravity arises from the fact that when one does various calculations—as, for instance, in attempting to calculate the probability that a gravitational wave will be scattered by another gravitational wave—one gets answers that turn out to be infinite. Another problem in the classical theory of gravitation arises from the presence of singularities: matter can apparently collapse to a point in space with infinite energy density and infinite space-time curvature. These absurdities, which have been exercising the attention of physicists for many decades, are precisely problems that involve gravity at very short distances—not the large distances of astronomy, but distances much smaller than the size of an atomic nucleus.

From the point of view of modern effective field theory, there are no infinities in the quantum theory of gravity. The infinities are cancelled in exactly the same way that they are in all our other theories, by just being absorbed into a redefinition of parameters in the field equations; but this works only if we include terms involving rates of fourth order and all higher orders. (John Donaghue of the University of Massachusetts at Amherst has done more than anyone in showing how this works.) The old problems of infinities and singularities in the theory of gravitation cannot be dealt with by taking Einstein's theory seriously as a fundamental theory. From the modern point of view—if you like, from my point of view— Einstein's theory is nothing but an approximation valid at long distances, which cannot be expected to deal successfully with infinities and singularities. Yet some professional quantum gravitationalists (if that is the word) spend their whole careers studying the applications of the original Einstein theory, the one that only involves second-order differential equations, to problems involving infinities and singularities. Elaborate formalisms have been developed that aim to look at Einstein's theory in a more sophisticated way, in the hope that doing so will somehow or other make the infinities or singularities go away. This ill-placed loyalty to general relativity in its original form persists because of the enormous prestige the theory earned from its historic successes.

But it is precisely in this way that the great heroic ideas of the past can weigh upon us, preventing us from seeing things in a fresh light. Said another way, it is those ideas that were most successful of which we should be most wary. Otherwise we become like the French army, which in 1914 tried to imitate the successes of Napoleon and almost lost the war—and then in 1940 tried to imitate the 1916 success of Marshall Petain in defending Verdun, only to suffer decisive defeat. Such examples exist in the history of physics as well. For instance, there is an approach to quantum field theory called second quantization, which fortunately no longer plays a significant role in research but continues to play a role in the way that textbooks are written. Second quantization goes back to a paper written in 1927 by Jordan and Klein that put forth the idea that after one has introduced a wave function in quantizing a theory of particles, you should then quantize the wave function. Surprisingly, many people think that this is the way to look at quantum field theory; it is not.

We have to expect the same fate for our present theories. The standard model of weak, electromagnetic, and strong forces, used to describe nature under conditions that can be explored in today's accelerators, may itself neither disappear nor be proved wrong but instead be looked at in quite a different way. Most particle physicists now think of the standard model as only an effective field theory that provides a low-energy approximation of a more fundamental theory.

Enough about the danger of history to science; let us now take up the danger of scientific knowledge to history. This arises from a tendency to imagine that discoveries are made according to our present understandings. Gerald Holton has done as much as anyone in trying to point out these dangers and puncture these misapprehensions. In his papers about Einstein he shows, for example, that the natural deduction of the special theory of relativity from the experiment of Michelson and Morley, which demonstrated that there is no motion through the ether, is not at all the way Einstein actually came to special relativity. Holton has also written about Kepler. At one point in my life I was one of those people who thought that Kepler deduced his famous three laws of planetary motion by studying the data of Tycho Brahe. But Holton points out how much else besides data, how much of the spirit of the Middle Ages and of the Greek world, went into Kepler's thinking—how many things that we now no longer associate with planetary motion were on Kepler's mind. By assuming that scientists of the past thought about things the way we do, we make mistakes; what is worse, we lose appreciation for the difficulties, for the intellectual challenges, that they faced.

Once, at the Tate Gallery in London, I heard a lecturer talking to a tour group about the Turner paintings. Turner was very important, said the guide, because he foreshadowed the Impressionists of the later nineteenth century. I had thought Turner was an important painter because he painted beautiful pictures; Turner did not know that he was foreshadowing anything. One has to look at things as they really were in their own time. This also applies, of course, to political history. Consider the term "Whig interpretation of history," which was invented by Herbert Butterfield in a lecture in 1931. As Butterfield explained it, "The Whig historian seems to believe that there is an unfolding logic in history." He went on to attack the person he regarded as the archetypal Whig historian, Lord Acton, who wished to use history as a way to pass moral judgments on the past. Acton wanted history to serve as the "arbiter of controversy, the upholder of that moral standard which the powers of earth and religion itself tend constantly to depress. . . . It is the office of historical science to maintain morality as the sole impartial criterion of men and things." Butterfield went on to say:

"If history can do anything it is to remind of us of those complications that undermine our certainties, and to show us that all our judgments are merely relative to time and circumstance. . . . We can never assert that history has proved any man right in the long run. We can never say that the ultimate issue, the succeeding course of events, or the lapse of time have proved that Luther was right against the pope or that Pitt was wrong against Charles James Fox."[2]

This is the point at which the historian of science and the historian of politics should part company. The passage of time has shown that, for example, Darwin was right against Lamarck, the atomists were right against Ernst Mach, and Einstein was right against the experimentalist Walter Kaufman, who presented data contradicting special relativity. To put it another way, Butterfield was correct; there is no sense in which Whig morality (much less the Whig Party) existed at the time of Luther. But nevertheless it is true that natural selection was working during the time of Lamarck, and the atom did exist in the days of Mach, and fast electrons behaved according to the laws of relativity even before Einstein. Present scientific knowledge has the potentiality of being relevant in the history of science in the way that the present moral and political judgements may not be relevant in political or social history.

Many historians, sociologists, and philosophers of science have taken the desire for historicism, the worry about falling into a Whig interpretation of history, to extremes. To quote Holton,

"Much of the recent philosophical literature claims that science merely staggers from one fashion, conversion, revolution, or incommensurable exemplar to the next in a kind of perpetual, senseless Brownian motion, without discernible direction or goal."[3] I made a similar observation in an address to the American Academy of Arts and Sciences about a year and a half ago, noting in passing that there are people who see scientific theories as nothing but social constructions. The talk was circulated by the Academy, as is their practice, and a copy of it fell into the hands of someone who over twenty years ago had been closely associated with a development known as the Sociology of Scientific Knowledge (SSK). He wrote me a long and unhappy letter; among other things, he complained about my remark that the Strong Program initiated at the University of Edinburgh embodied a radical social-constructivist view, in which scientific theories are nothing but social constructions. He sent me a weighty pile of essays, saying that they demonstrated that he and his colleagues do recognize that reality plays a role in our world. I took this criticism to heart and decided that I would read the essays. I also looked back over some old correspondence that I had had with Harry Collins, who for many years led the well-known Sociology of Scientific Knowledge group at the University of Bath. My purpose in all of this was to look at these materials from as sympathetic a point of view as I could, try to understand what they were saying, and assume that they must be saying something that is not absurd.

I did find described (though not espoused) in an article by David Bloor, who is one of the Edinburgh group, and also in my correspondence with Harry Collins, a point of view that on the face of it is not absurd. As I understand it, there is a position called "methodological idealism" or "methodological antirealism," which holds that historians or sociologists should take no position on what is ultimately true or real. Instead of using today's scientific knowledge as a guiding principle for their work, the argument goes, they should try to look at nature as it must have been viewed by the scientists under study at the time that those scientists were working. In itself, this is not an absurd position. In particular, it can help to guard us against the kind of silliness that (for instance) I was guilty of when I interpreted

Kepler's work in terms of what we now know about planetary motion.

Even so, the attitude of methodological antirealism bothered me, though for a while I could not point to what I found wrong with it. In preparing this essay I have tried to think this through, and I have come to the conclusion that there are a number of minor things wrong with methodological antirealism: it can cripple historical research, it is often boring, and it is basically impossible. More significantly, however, it has a major drawback—in an almost literal sense, it misses the point of the history of science.

Let us first address the minor points. If it were really possible to reconstruct everything that happened during some past scientific discovery, then it might be helpful to forget everything that has happened since; but in fact much of what occurred will always be unknown to us. Consider just one example. J. J. Thomson, in the experiments that made him known as the discoverer of the electron, was measuring a certain crucial quantity, the ratio of the electron's mass to its charge. As always happens, he found a range of values. Although he quoted various values in his published work, the values he would always refer to as his favorite results were those at the high end of the range. Why did Thomson quote the high values as his favorite values? It is possible that Thomson knew that on the days those results had been obtained he had been more careful; perhaps those were the days he had not bumped into the laboratory table, or before which he'd had a good night's sleep. But the possibility also exists that perhaps his first values had been at the high end of the range, and he was determined to show that he had been right at the beginning. Which explanation is correct? There is simply no way of reconstructing the past. Not his notebooks, not his biography—nothing will allow us now to reconstitute those days in the Cavendish Laboratory and find out on which days Thomson was more clumsy or felt more sleepy than usual. There is one thing that we do know, however: the actual value of the ratio of the electron's mass to its charge, which was the same in Thomson's time as in our own. We know, in fact, that the actual value is not at the high end but, rather, at the low end of the range of Thomson's experimental values, which strongly suggests that when Thomson's measurements gave high values they were not actually more careful—and that therefore it is more likely that Thomson quoted these values because he was just trying to justify his first measurements.

This is a trivial example of the use of present scientific knowledge in the history of science, because here we are just talking about a number, not a natural law or an ontological principle. I chose this example simply because it shows so clearly that to decide to ignore present scientific knowledge is often to throw away a valuable historical tool.

A second minor drawback of methodological antirealism is that a reader who does not know anything about our present understanding of nature is likely to find the history of science terribly boring. For instance, a historian might describe how in 1911 the Dutch physicist Kamerlingh Onnes was measuring the electrical resistance of a sample of cold mercury and thought that he had found a short circuit. The historian could go on for pages and pages describing how Onnes searched for the short circuit, and how he took apart the wiring and put it back together again without any success in finding the source of the short circuit. Could anything be more boring than to read this description if one did not know in advance that there *was* no short circuit—that what Onnes was observing was in fact the vanishing of the resistance of mercury when cooled to a certain temperature, and that this was nothing less than the discovery of superconductivity? Of course, it is impossible today for a physicist or a historian of physics not to know about superconductivity. Indeed, we are quite incapable while reading about the experiments of Kamerlingh Onnes of imagining that his problem was nothing but a short circuit. Even if one had never heard of superconductivity, the reader would know that there was something going on besides a short circuit; why else would the historian bother with these experiments? Plenty of experimental physicists have found short circuits, and no one studies them.

But these are minor issues. The main drawback of methodological antirealism is that it misses the point about the history of science that makes it different from other kinds of history: Even though a scientific theory is in a sense a social consensus, it is unlike any other sort of consensus in that it is culture-free and permanent.

This is just what many sociologists of science deny. David Bloor stated in a talk at Berkeley a year ago that "the important thing is that reality underdetermines the scientists' understanding." I gather he means that although he recognizes that reality has some effect on what scientists do—so that scientific theories are not "nothing but" social constructions—scientific theories are also not what they are simply because that is the way nature is. In a similar spirit, Stanley Fish, in a recent article in the *New York Times*, argued that the laws of physics are like the rules of baseball. Both are certainly conditioned by external reality—after all, if baseballs moved differently under the influence of Earth's gravity, the rules would call for the bases to be closer together or further apart—but the rules of baseball also reflect the way that the game developed historically and the preferences of players and fans.[4]

Now, what Bloor and Fish say about the laws of nature does apply while these laws are being discovered. Holton's work on Einstein, Kepler, and superconductivity has shown that many cultural and psychological influences enter into scientific work. But the laws of nature are not like the rules of baseball. They are culture-free and they are permanent—not as they are being developed, not as they were in the mind of the scientist who first discovers them, not in the course of what Latour and Woolgar call "negotiations" over what theory is going to be accepted, but in their final form, in which cultural influences are refined away. I will even use the dangerous words "nothing but": aside from inessentials like the mathematical notation we use, the laws of physics as we understand them now are nothing but a description of reality.

I cannot prove that the laws of physics in their mature form are culture-free. Physicists live embedded in the Western culture of the late twentieth century, and it is natural to be skeptical if we say that our understanding of Maxwell's equations, quantum mechanics, relativity, or the standard model of elementary particles is culture-free. I am convinced of this because the purely scientific arguments for these theories seem to me overwhelmingly convincing. I can add that as the typical background of physicists has changed, in particular, as the number of women and Asians in physics has increased, the nature of our understanding of physics has not changed. These laws in their mature form have a toughness that resists cultural influence.

The history of science is further distinguished from political or artistic history (in such a way as to reinforce my remarks about the influence of culture) in that the achievements of science become permanent. This assertion may seem to contradict a statement at the beginning of this essay—that we now look at general relativity in a different way than Einstein did, and that even now we are beginning to look at the standard model differently than we did when it was first being developed. But what changes is our understanding of both why the theories are true and their scope of validity. For instance, at one time we thought there was an exact symmetry in nature between left and right, but then it was discovered that this is only true in certain contexts and to a certain degree of approximation. But the symmetry between right and left was not a simple mistake, nor has it been abandoned; we simply understand it better. Within its scope of validity, this symmetry has become a permanent part of science, and I cannot see that this will ever change.

In holding that the social constructivists missed the point, I have in mind a phenomenon known in mathematical physics as the approach to a fixed point. Various problems in physics deal with motion in some sort of space. Such problems are governed by equations dictating that wherever one starts in the space, one always winds up at the same point, known as a fixed point. Ancient geographers had something similar in mind when they said that all roads led to Rome. Physical theories are like fixed points, toward which we are attracted. Starting points may be culturally determined, paths may be affected by personal philosophies, but the fixed point is there nonetheless. It is something toward which any physical theory moves; when we get there we know it, and then we stop.

The kind of physics I have done for most of my life, working in the theory of fields and elementary particles, is moving toward a fixed point. But this fixed point is unlike any other in science. That final theory toward which we are moving will be a theory of unrestricted validity, a theory applicable to all phenomena throughout the universe—a theory that, when finally reached, will be a permanent part of our knowledge of the world. Then our work as elementary particle physicists will be done, and will become nothing but history.

ENDNOTES

[1]Steven Weinberg, *The Discovery of Subatomic Particles* (San Francisco: Scientific American/Freeman, 1982).

[2]Herbert Butterfield, *The Whig Interpretation of History* (New York: Scribners, 1951), 75.

[3]Gerald Holton, *Einstein, History, and Other Passions* (Reading, Mass.: Addison-Wesley, 1996), 22.

[4]Stanley Fish, "Professor Sokal's Bad Joke," *New York Times*, 21 May 1996, Op-Ed section.

Bretislav Friedrich and Dudley Herschbach

Space Quantization:
Otto Stern's Lucky Star

*Much of my work has had its origin in the notion
that science should treasure its own history, that
historical scholarship should treasure science, and
that the full understanding of each is deficient
without the other.*

*—Gerald Holton
The Advancement of Science, and Its Burdens[1]*

IN THIS ESSAY WE REVISIT A TREASURED EPISODE from the heroic
age of atomic physics. The story centers on an experiment,
elegantly simple in its conception, extraordinarily startling in
its outcome, and extremely fruitful in its legacy. From it emerged
both new intellectual vistas and a host of useful applications of
quantum science. Yet this germinal experiment, carried out at
Frankfurt in 1921–22 by Otto Stern and Walther Gerlach, is not
at all familiar except to physical scientists.[2] Even among them
we have found very few aware of historical particulars that
enhance the drama of the story and the abiding lessons it offers
about how science works. These particulars include a bad cigar
that amplified a puny signal, a postcard from New York that
offset the huge inflation then rising in Germany, and an uncanny
"conspiracy of Nature" that rewarded the audacity of the experimenters, despite the inadequacies of a fledgling theory that
had led a skeptical Otto Stern to devise his crucial test.

*Bretislav Friedrich is Senior Research Fellow in the Department of Chemistry and
Chemical Biology at Harvard University.*

Dudley Herschbach is the Baird Professor of Science at Harvard University.

165

We begin by describing the historical context of the experiment, chiefly stemming from the atomic model proposed by Niels Bohr in 1913, nowadays referred to as the "old quantum theory." Our intent is to provide an account accessible to anyone with only vague memories of high-school physics or chemistry. But as background we need to discuss a few concepts, to show how Stern's interest was whetted by the tantalizing, partial successes and patent failures of Bohr's model when confronted with atomic spectra and magnetism. Stern came to focus on the idea of *space quantization.* This was one of the most peculiar inferences from the old quantum theory, the notion that the quasiplanetary electron orbits postulated by Bohr could have only certain discrete orientations in space. Even Stern's theoretical colleagues who had invoked this idea regarded it as merely a mathematical construct, devoid of physical reality.

We next trace the conception, preparation, vicissitudes, and realization of the Stern-Gerlach experiment. It showed unequivocally that space quantization was real, and thus provided compelling evidence that a new mechanics was required to describe the atomic world. Myriad experiments since have confirmed and exploited space quantization. Ironically, however, the seeming agreement of the original experiment with the old quantum theory was chimerical. Within a few years, the electron orbits of the Bohr model were shown not to exist. Another electronic property was discovered, called spin, that produced an equivalent result in the Stern-Gerlach experiment. Yet space quantization was thereby reincarnated, in a more comprehensive and comprehensible form. As an epilogue, we briefly describe the modern incarnation and inspect treasures it has minted.

PRELUDE: OTTO STERN AND THE BOHR ATOM

Otto Stern (1888–1969) received his Ph.D. in 1912 at Breslau in physical chemistry.[3] His doctoral dissertation presented theory and experiments on concentrated solutions of carbon dioxide in various solvents, just generalized soda water. His parents, proud and affluent, offered to support him for postdoctoral study anywhere he liked. Motivated by "a spirit of adventure," Stern opted to work with Einstein, then at Prague. They had not met, but Stern knew Einstein was "a great man, at the center of modern developments." Contact was quickly made via what would now be called "the old-boy network" and Einstein (then thirty-three) indicated he was willing to accept Stern (then twenty-four). In Prague, Einstein held discussions "with his first pupil, Otto Stern, . . . in a café which was attached to a brothel."[4] Soon Einstein was recalled to Zurich; Stern accompanied him there and in 1913 was appointed *privatdozent* for physical chemistry.

Under Einstein's influence, Stern became interested in light quanta, the nature of atoms, magnetism, and statistical physics. However, Stern was shocked when in mid-1913 Bohr published his iconoclastic atomic model. Soon after, Stern discussed it thoroughly with his colleague Max von Laue during a long walk up the Ütliberg, a mountain near Zurich. This led them to make a solemn oath that later acquired some notoriety: "If this nonsense of Bohr should in the end prove to be right, we will quit physics."[5]

The quantization of energy had first been boldly invoked by Max Planck in 1900 and by Einstein in 1905, but they dealt with many-particle systems. For the most part, the physics community had suspended judgment, supposing that some way might yet be found to reconcile seemingly aberrant phenomena with the concepts of classical mechanics and electromagnetic theory that were so securely established in macroscopic physics. Bohr's work was more perplexing and immediate in its impact, as he treated the simplest atom, hydrogen—comprised of just two particles, a positively charged proton and an negatively charged electron. Ernest Rutherford had shown in 1911 that nearly all of the mass but only a minuscule fraction of the volume of an atom resided in its nucleus (here, the proton). That suggested a model for the atom similar to the solar system, with a planetary electron circling the nucleus. The problem was that, according to classical physics, such an atom would collapse in an instant. The electrical attraction to the proton would cause the electron to spiral rapidly into the nucleus, giving up its kinetic energy as radiation.

"A Triumph over Logic"

Bohr simply postulated that the electron could circle the proton only on one or another of a discrete set of orbits. These he called "stationary states." He offered no justification for such a blatant violation of classical mechanics, which would permit a continuous range of possible orbits. He also proposed that the electron could switch inward or outward, from one orbit to another, by emitting or absorbing a quantum of light, with wavelength determined by the difference in energy of the initial and final orbits. But he asserted that somehow the innermost orbit, closest to the nucleus, was stable; from this "ground state" the electron would not fall into the nucleus. Finally, Bohr came up with a means to calculate the size of the orbits and their energies. It amounted to postulating that, in addition to the energy, the angular momentum of the electron was quantized. For a circular orbit, classical mechanics defines the angular momentum L as a vector, perpendicular to the orbit, with magnitude given by the product of the electron's mass, its velocity, and the orbital radius. Again, in classical physics L could have a continuous range of values, whereas the quantization condition adopted by Bohr specified $L = n(h/2\pi)$. Here $n = 1, 2, 3 \ldots$ is an integer (with $n = 1$ for the ground state) and h is Planck's fundamental constant, the proportionality factor between frequency and energy that appeared in the quantization rule Planck and Einstein had employed in entirely different contexts than Bohr's atomic model.

In the apt phrase of Abraham Pais, the weird model concocted by Bohr proved a "triumph over logic."[6] It scored a stunning success in accounting for major features of the observed spectrum of the hydrogen atom. In 1885 (the year of Bohr's birth), a remarkably simple empirical formula known as the Balmer formula had been found, which related frequencies (or reciprocal wavelengths) of the spectral lines to the differences of reciprocal squares of integer numbers. This relationship had remained an unexplained curiosity. Bohr's calculations gave an expression of the same functional form, in which the integers involved were simply the values of his quantum number n for the initial and final orbits involved in the electronic transition. Moreover, Bohr was able to evaluate a proportionality factor in the Balmer formula, known as the Rydberg constant, in terms of the charge and mass of the electron and Planck's constant. These fundamental quantities were not yet accurately known, so his theoretical value was uncertain by a few percent; but Bohr's result agreed within that range with the empirical value of the Rydberg constant.

Likewise, Bohr's quantization of angular momentum enabled him to calculate the radius of an orbit in terms of the electronic charge, the mass, and Planck's constant. Again, he found satisfactory agreement with empirical estimates of the atomic size. In this case the comparison was merely in order of magnitude, yet quite significant since other extant models provided no means to predict the radius of an atom. In fact, the best empirical estimates then available came from diffraction of x-rays by crystal lattices—a method invented in 1912 by none other than von Laue, who thereby helped provide evidence supporting a model he found exceedingly distasteful.

A corollary of Bohr's model for the hydrogen atom had a particularly compelling success. In 1896 Charles Pickering, a Harvard astronomer, had discovered in starlight another remarkably regular series of spectral lines, one not seen in laboratory spectra of hydrogen but likewise involving differences of reciprocal squares of integers. Bohr noted that these unassigned lines could be ascribed to the helium atomic ion; like hydrogen, it has only one electron, but the helium nucleus contains two protons. Accordingly, Bohr's model predicted that the Balmer formula should apply with the Rydberg constant increased by 2^2, or 4. This nicely accounted for the spectral pattern, but a spectroscopist then pointed out that the factor would need to be 4.0016 rather than 4 to fit the lines accurately. Bohr responded that, for simplicity, he had previously approximated the mass of the nucleus as infinitely heavier than the electron; the correction could be easily and precisely evaluated since it depended only on accurately known mass ratios. He thus obtained a factor of 4.00163, in gratifying and unprecedented quantitative agreement with the data. Einstein commented that this was "an enormous achievement."

Despite these and other happy results, Bohr, like many of his contemporaries, was very dissatisfied with his model. He regarded it as "makeshift and provisional," not only because it was in "conflict with the admirably coherent conceptions which have been rightly termed the classical theory of electrodynamics," but because it "has too much of approximation in it and it is philosophically not right." Its inadequacies became glaringly evident also in attempts to account for further experimental results. Efforts to extend Bohr's approach to calculate the spectra of two-electron atoms, such as unionized helium, failed dismally. Another recalcitrant puzzle was the splitting of lines caused by the application of an external magnetic field, known as the Zeeman effect. We will consider this specifically, since Stern was to exploit atomic magnetism. However, first we need to describe an extension of Bohr's model, revealing the aspect destined to be confirmed by Stern as a key element of nature's strange logic for the atomic world.

Space Quantization

Even the hydrogen atom posed more puzzles, clues for deeper issues. At high resolution, the spectral lines had a fine structure, which was not included in the Balmer formula. This structure, as well as the shifts and splittings induced by application of external or magnetic fields, spurred efforts to develop a more comprehensive theory. In 1916, Arnold Sommerfeld and, independently, Peter Debye generalized Bohr's model for hydrogen. They introduced three quantization conditions, pertaining to different components of the orbital angular momentum of the electron. The allowed discrete orbits of the electron were then not limited to circles but were in general ellipses, characterized by three quantum numbers.

One of these, $n = 1, 2, 3, \ldots$, now called the principal quantum number, was like that which Bohr had invoked for his circular orbits. The two new quantum numbers, denoted by k and m, were also restricted to integer values. The quantity k, termed the azimuthal quantum number, ranges from $k = 1$, $2, \ldots, n$; together with n, the value of k determines the *size* and *shape* of the elliptical orbit (which becomes circular for $k = n$). The quantity m, called the projection quantum number, ranges

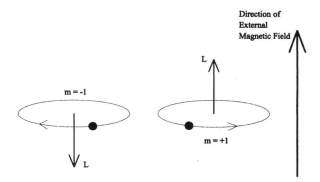

Figure 1: Space Quantization According to the Old Quantum Theory. Depicted is the simplest case: a single electron orbiting about the atomic nucleus, with one quantum of angular momentum. The angular momentum vector L (and the atomic magnetic moment which is proportional to it) has only two allowed orientations, either parallel (projection quantum number *m* = 1) or antiparallel (*m* = -1) to the direction of an external magnetic field.

from -*k* to +*k* in integer steps (but with *m* = 0 excluded); it specifies the *orientation* of the plane of the elliptical orbit in space. Equivalently, since the orbital angular momentum vector L is perpendicular to the plane of the orbit, *m* also determines the projection of L on any prescribed axis. These limitations to discrete orientations, corresponding to the integer values of *m*, were termed directional or spatial or *space quantization.*

Figure 1 pictures this model for the simplest case, the ground state of hydrogen. There *n* = 1 so only *k* = 1 is allowed and accordingly just *m* = -1 and + 1. Hence for this case the orbit is circular, but its angular momentum L can take up only two discrete orientations, corresponding to clockwise and counterclockwise motion of the electron. (We refrain from displaying the set of elliptical orbits for larger *n*, still a favorite decoration for textbooks.)[7]

Sommerfeld showed that, for his model, the energy was not affected by *k* or *m* but depended only on the principal quantum number *n*, just as in the original Bohr model, provided that relativistic effects were neglected and external electric or magnetic fields were absent. However, he also evaluated a small contribution to the energy, depending on both *n* and *k*, which arose from the relativistic change of the electron mass as its velocity varies in the elliptic orbit. His relativistic correction was found to give good quantitative agreement with the observed spectral fine structure; it was justly considered "a triumph both for quantum and for relativity theory."

However, in the absence of an external field, the role of spatial quantization remains unobservable. The energy and hence the spectrum of the atom are then independent of the m quantum number. That holds, as Sommerfeld noted, because the various discrete orientations specified by m are equivalent if there is no "preferred direction of reference in space." When such a special direction is imposed by an external electric or magnetic field, the various discrete orientations can differ in energy. That occurs simply because in general the interaction of an electron with the field depends on the angle between the field direction and the electron's path. Now we turn to some pertinent aspects of magnetism, to show how the actors and issues of our story became aligned in a fortunate direction.

Magnetism and the Zeeman Effect

Back in 1820, Hans Christian Ørsted, the leading Danish physicist of his day, discovered that an electric current generates a magnetic field. Soon after, André Ampère conjectured that magnetism in matter arose from charged particles moving in tiny circuits. Bohr, who had done his doctoral thesis on the electron theory of metals, took the opportunity—ninety years later—to compute the strength of the elementary magnet for his atomic model. For an electron in circular orbit with quantum number n, he found that the magnetic moment was n times a quantity now called the Bohr magneton; it is proportional to Planck's constant times the ratio of the electronic charge to its mass. Curiously, Bohr did not include that result in his 1913 papers, although it appears in an existing draft.

Ampère's conjecture also interested Einstein, who in 1915 with Wander de Haas even undertook an experiment to determine the ratio between the magnetic moment of electrons in iron and the angular momentum associated with the postulated electron orbits. Einstein remarked, "How tricky Nature is when one tries to approach her experimentally! In my old days [he was thirty-six] I am developing a passion for experiments."[8] The experimental result seemed to confirm a prediction derived from classical electromagnetic theory. Bohr cited this as confirming his postulate that electrons can circulate in atoms without emitting radiation. However, the experiment was trickier than Einstein realized, and so was the theory; not until more than a decade later was it known that the magnetism of iron comes almost solely from electron spins, not orbital motion.

Stern first became involved with magnetism while in Zurich, where in discussions he helped refine the theory of ferromagnetism advanced in 1913 by Pierre Weiss. That theory, still useful today, envisioned magnetic domains within a metal. However, it implied that the average magnetic moment of an atom in a fully magnetized sample of iron was much smaller than the Bohr magneton, by about a factor of five. In an attempt to account for this, in 1920 Wolfgang Pauli invoked the idea of space quantization, noting that the apparent magnetic moment an atom contributes within a domain depends on the atom's orientation with respect to the field direction. He performed a statistical average over the projection quantum numbers m and concluded the net effective atomic moment should indeed be much smaller than the Bohr magneton. Again, as with Einstein and de Haas, the basic model was wrong (since spin rather than orbital magnetism is involved). However, Pauli's appeal to space quantization of atomic magnets has historical significance in that it made his colleagues, including Stern, mindful of the idea.

Although the old quantum theory was not obviously in conflict with these studies of bulk magnetism, it had no such luck with the Zeeman effect. Soon after the splitting of spectral lines in a magnetic field was discovered in 1897, Hendrick Lorentz offered an explanation based on a classical model of the atom: depending on whether the emitted light was viewed parallel or perpendicular to the field direction, a line should split into a doublet or triplet, with spacings proportional to the field strength. Such behavior, which came to be termed the "normal" Zeeman effect, was found in a few cases, at least for weak fields. But that proved to be abnormal. Most often, lines split into more than three components, and the spacings were not simply proportional to the field strength; this typical situation was termed the "anomalous" Zeeman effect.

In the augmented Bohr model developed by Sommerfeld and by Debye, space quantization provided a nice explanation of the normal Zeeman effect. As noted above, when a field is present, the orbits with different spatial orientations differ in energy. Accordingly, if the projection quantum number m changes for an electron jump between orbits, the corresponding spectral line shifts from the position it had in the absence of the field, so the original line appears to split up. This success with the normal Zeeman effect could not be taken as evidence for space quantization, however, since even Lorentz's simple classical model appeared adequate for the normal case. Furthermore, despite strenuous efforts by Sommerfeld, Debye, and others, no way was found to account for the complexities of the anomalous Zeeman effect. Thus, the notion of space quantization did not enable the old quantum theory to do any better than the classical theory in coping with the Zeeman effect.

As the quantity and quality of spectroscopic data grew, the intractable anomalous effect spread bafflement and gloom. Here is a lament by Pauli:

> The anomalous type of [magnetic] splitting . . . was hardly understandable, since very general assumptions concerning the electron, using classical theory as well as quantum theory, always led to the same triplet. A closer investigation of this problem left me with the feeling that it was even more unapproachable. A colleague who met me strolling rather aimlessly in the beautiful streets of Copenhagen said to me in a friendly manner, "You look very unhappy," whereupon I answered fiercely, "How can one look happy when he is thinking of the anomalous Zeeman effect?"[9]

Thus it came to pass that atomic spectra, which had provided much encouragement for the fledgling quantum theory, also revealed most clearly its inadequacies. Both proponents and critics were stymied. Then, in 1921, Otto Stern proposed a definitive experiment, not involving spectroscopy. He asserted that "the experiment, if successful, will decide unequivocally between the quantum theoretical and classical views" and would thereby prove whether or not space quantization exists.[10]

THE STERN-GERLACH EXPERIMENT

The immediate stimulus for Stern was a property implied by the Sommerfeld-Debye theory that had *not* been observed. According to the theory, as illustrated in figure 1, hydrogen atoms (with $n = 1$) in a magnetic field would be aligned such that the electron orbits are perpendicular to the direction of the field ($m = \pm 1$). A beam of light directed perpendicularly to the magnetic field would then interact differently with the orbiting electron, depending on whether the electric vector of the light oscillates parallel or perpendicular to the magnetic field. In the parallel case, the oscillating electric vector of the light acts to pull the electron out of its orbital plane; in the perpendicular case, it would displace the electron in the orbital plane. The propagation velocity of the light through a gas of hydrogen atoms, and hence the index of refraction, should therefore differ markedly for the parallel and perpendicular cases.

As the same considerations apply for many-electron atoms or molecules, the old quantum theory predicted that any gas should be expected to exhibit birefringence, a phenomenon well known in optics of anisotropic liquid and solid media. However, magnetically induced birefringence of gases had never been observed. This cast yet another shadow on the old quantum theory but awakened in Stern an illuminating insight. As he recalled the creative moment:

> The question whether a gas might be magnetically birefringent was raised at a seminar. The next morning I awoke early, too early to go to the lab. As it was too cold to get out of bed, I lay there thinking about the seminar question and had the idea for the experiment.[11]

Stern's key idea was to look for space quantization by using the magnetism of the atom as a probe. If space quantization occurs, the atomic magnets would have only discrete projections (specified by the quantum number m) on the direction of an external magnetic field. In contrast, according to classical mechanics, as long as the atoms do not undergo collisions, the atomic magnets would remain randomly oriented whether or not an external field is present. By conceiving this *Gedankenexperiment*, in Einstein's style, Stern showed how to decide the issue.

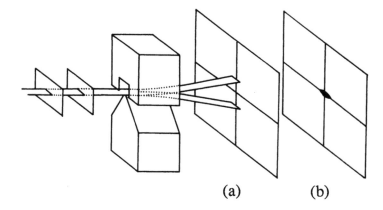

(a)   (b)

Figure 2: Schematic View of the Stern-Gerlach Apparatus. Indicated are (a) the observed beam splitting and (b) the unsplit outcome predicted by classical mechanics. The beam of silver atoms, produced by effusion of metallic vapor from an oven heated to 1000°C, is collimated by two slits (0.03 millimeters wide in the vertical direction). The beam passes through an inhomogeneous magnetic field about 3.5 centimeters long; its direction and gradient are vertical, the field strength is about 0.1 Tesla and the gradient 10 Tesla/cm. The transmitted beam is deposited on a cold glass plate. The magnitude of the splitting is only 0.2 millimeters. For the sake of clarity, in the figure the splitting is much exaggerated and the broadening of the deposits caused by the thermal distribution of velocities in the beam is omitted.

Plans and Preparations

Stern envisaged a well-collimated beam of silver atoms, traveling in a vacuum chamber and passing through a magnetic field directed across the beam path. Since a silver atom has only one valence electron, for his purpose it could be expected to behave like hydrogen (which is less convenient to handle experimentally). The beam is so dilute that the individual atoms sail through the apparatus without interacting with others. After traveling between the pole pieces of the magnet, the atoms land on a cold glass plate to which they adhere and thereby exhibit the beam intensity profile. As pictured in figure 2, one of the magnet pole pieces has a sharp edge, the other a broad notch; this makes the magnetic field stronger near the edge, weaker near the notch. In this nonuniform field, transverse to the beam path, an atom is subject to a deflecting force proportional to the angle between the atomic magnet and the external field gradient.

Consequently, the atomic magnets that are tilted towards the field direction are attracted to the stronger field region, whereas those tilted away are repelled. The trajectories of atoms emerging from the deflecting magnet, as recorded by deposits on the glass plate, thus reveal the spatial orientation of the atomic magnets.

With such a setup, Stern predicted that space quantization would produce a *splitting* of the atomic beam into two distinct components, since in the ground state of the silver atom the valence electron was expected to have just one unit of orbital angular momentum ($n = k = 1$, so $m = +1$ and -1 components). For any classical model, however, the atomic magnets would be distributed over a continuous angular range, so passing through the deflecting field would not split the beam but only *broaden* it along the field direction.

Stern was at this time assistant to Max Born at the Institute for Theoretical Physics in Frankfurt. Soon after hatching his idea in a warm bed, Stern hastened to Born, who gave it a cool assessment. As Born recalled:

> It took me quite a time before I took this idea seriously. I thought always that direction [space] quantization was a kind of symbolic expression for something which you don't understand. But to take this literally like Stern did, this was his own idea. . . . I tried to persuade Stern that there was no sense [in it], but then he told me that it was worth a try.[12]

Stern was, with Born's blessing, already engaged in an experiment to test a central theoretical result of statistical physics, the form of the distribution of molecular velocities in a gas. It had been derived decades earlier by James Clerk Maxwell and Ludwig Boltzmann, but Stern's work was the first direct test. He employed a beam of silver atoms, effusing from a small oven in vacuum, and scanned the velocity distribution by observing the transmission of the beam through a slit system rotating at high speed. Born described how Stern conducted this work:

> I had only two rooms in Frankfurt. . . . Stern's apparatus was made up in my little room, so I saw it from the beginning and watched. And I was quite envious of how he managed: he did not touch it at all, for he is also, just like me, not very good with his

hands. But we had a very good mechanic.... [Stern] told him what to do and it came out.[13]

The experimental results proved to be in agreement with the Maxwell-Boltzmann distribution, after Einstein provided some help with the interpretation.

Adjacent to Born's Theoretical Institute was that for Experimental Physics, where Walther Gerlach was a newly hired member. He was reputed to be an excellent experimentalist, and moreover had undertaken work with atomic beams. He wanted to determine whether a beam of bismuth atoms was magnetic, in contrast to solid bismuth, which is not. He planned to do this by sending the beam through the same sort of magnetic deflecting field that Stern had in mind. Stern promptly recruited Gerlach, saying: "With the magnetic [deflection] experiment one can do something else. Do you know what directional [space] quantization is?" Gerlach did not. After a brisk, excited explanation, Stern concluded with: "Shall we do it? Well, let's go, we shall do it!"[14]

Realization and Reception

Stern's design calculations indicated the experiment was barely feasible; indeed, despite the simplicity of his scheme, it took more than a year to accomplish. The apparatus had two vacuum chambers—one held the oven that produced the beam of silver atoms, the other contained an electromagnet and the glass collector plate. The beam collimation had to be extremely narrow if the small splitting were to be resolved, so the beam intensity at the collector plate was very low. The attainable "exposure time" was usually only a few hours, between breakdowns of the apparatus. Thus only a meager film of silver atoms was deposited, too thin to be visible to an unaided eye. Forty years later, Stern enjoyed recalling a cherished episode:

> After venting to release the vacuum, Gerlach removed the detector flange. But he could see no trace of the silver atom beam and handed the flange to me. With Gerlach looking over my shoulder as I peered closely at the plate, we were surprised to see gradually emerge the trace of the beam.... Finally we realized what [had happened]. I was then the equivalent of an assistant professor.

My salary was too low to afford good cigars, so I smoked bad cigars. These had a lot of sulfur in them, so my breath on the plate turned the silver into silver sulfide, which is jet black, so easily visible. It was like developing a photographic film.[15]

After this, Gerlach and Stern began using a photographic development process. However, other devilish difficulties persisted. As inconclusive efforts continued for months, Stern's assessment of space quantization wavered back and forth, between conviction and rejection. Gerlach's faith was also being undermined by dubious colleagues, including Debye: "But surely you don't believe that the [spatial] orientation of atoms is something physically real; that is [only] a prescription for the calculation, a timetable for the electrons."[16]

During this gestatory period, Stern left Frankfurt to assume the post of professor of theoretical physics at Rostock, returning during vacations to work on the experiment. Its formidable character and the fortitude of Gerlach has been vividly described by one of his students:

Anyone who has not been through it cannot at all imagine how great were the difficulties with an oven to heat the silver up ... within an apparatus which could not be fully heated [the seals would melt] and where a vacuum ... had to be produced and maintained for several hours. The pumping speed ... was ridiculously small compared with the performance of modern pumps. And ... the pumps were made of glass and quite often they broke, either from the thrust of boiling mercury ... or from the dripping of condensed water vapor. In that case the several-day effort of pumping, required during the warming up and heating of the oven, was lost. Also, one could be by no means certain that the oven would not burn through during the four- to eight-hour exposure time. Then both the pumping and the heating of the oven had to be started from scratch. It was a Sisyphus-like labor and the main load and responsibility was carried on the broad shoulders of Professor Gerlach. ... He would get in about 9 p.m. equipped with a pile of reprints and books. During the night he then read the proofs and reviews, wrote papers, prepared lectures, drank plenty of cocoa or tea and smoked a lot. When I arrived the next day at the Institute, heard the intimately familiar noise of the running pumps, and found Gerlach still in the lab, it was a good sign: nothing broke during the night.[17]

When it was ultimately resolved, the observed splitting of the silver beam was only 0.2 millimeters. Accordingly, a misalignment of the oven orifice, the pair of collimation slits, or the edge of the "sharp" pole piece of the magnet by more than 0.01 mm was enough to spoil an experimental run.

Another handicap was the financial disarray that began to beset Germany in 1920. Born tells about it:

> We were already in the inflation which later became so disastrous; but we were not aware of what was happening. Everything was scarce and expensive. Physical instruments were hardly obtainable. So my funds were quickly exhausted. . . . At that time a wave of interest in Einstein and his theory of relativity was sweeping the world. He had predicted the deflection, by the sun, of light coming from a star . . . after laborious measurements and tedious calculations the conclusion was arrived at [in 1919] that Einstein was right, and this was published under sensational headlines in all the newspapers. . . . There was an Einstein craze, everybody wanted to learn what it was all about. . . . I announced a series of three lectures in the biggest lecture-hall of the University on Einstein's theory of relativity and charged an entrance fee for my Department. It was a colossal success, the hall was crowded and a considerable sum collected. . . . The money thus earned helped us for some months, but as inflation got worse, it evaporated quickly and new means had to be found.

> One day I met a friend . . . who was going to New York. . . . I said jokingly: "If you find a German-American who is still interested in the old country, tell him I need dollars for important experiments in my Department." I had quite forgotten this remark when a few weeks later a postcard arrived: . . . "Write to Henry Goldman [of Goldman Sachs, and also the progenitor of Woolworth stores], 998 Fifth Avenue, New York." At first I took it for another joke, but on reflection I decided that an attempt should be made. . . . [A] nice letter was composed and despatched, and soon a most charming reply arrived and a cheque for some hundreds of dollars which helped us out of our difficulties. . . . After Goldman's cheque had saved our experiments, the work [on the Stern-Gerlach experiment] went on successfully.[18]

Einstein himself also helped. He was then the director of the Kaiser-Wilhelm Institute of Physics in Berlin, and provided a grant from the endowment of his Institute.

When Gerlach at last did clearly resolve the beam splitting, he informed Stern, then at Rostock, by telegram.[19] Ironically, just then Stern's doubts about space quantization were again ascendant; later, he recalled the surprise and excitement as overwhelming. Despite the small size of the splitting, from careful analysis Gerlach and Stern were even able to determine the magnetic moment of the silver atom. They found it equal to the Bohr magneton (within an accuracy of about 10 percent)—in gratifying agreement with the result expected from the old quantum theory, if the magnetism arose from orbital motion of the valence electron with one quantum of angular momentum.

The directness and conceptual simplicity of the Stern-Gerlach experiment ensured that it had great impact.[20] It was immediately accepted as among the most compelling evidence for quantum theory. But space quantization was a double-edged discovery. Einstein and Ehrenfest, among others, struggled without success to understand how the atomic magnets could take up definite, preordained orientations in the field. Likewise, the lack of birefrigence became a more insistent puzzle. Those questions and others (such as the anomalous Zeeman effect) could not be cleared up for a few more years, until further discoveries ushered in modern quantum mechanics.

EPILOGUE: AN ABIDING LEGACY

Having visited the Stern-Gerlach experiment in its historical context, rife with hesitant and confusing theoretical ideas, we now look briefly at the interpretation that has become canonical. This involves no less than four major pillars of quantum physics that emerged during the years from 1925 to 1927: electron spin, a deeper view of angular momentum, wave-particle duality, and the uncertainty principle. These discoveries made obsolete the old quantum theory, but enhanced the scope and significance of space quantization.

Atomic spectra, burgeoning with fine structure and Zeeman splittings, provoked theorists to resort to attempting empirical schemes that postulated a variety of ad hoc angular momenta and quantization rules. After curious twists and turns, this led George Uhlenbeck and Samuel Goudsmit, young graduate students at Leiden, to propose in 1925 that an electron has an intrinsic angular momentum or "spin."[21] Its allowed projections (in units of $h/2\pi$) are only $m_s = \pm 1/2$, in contrast to the integer values that occur for orbital angular momentum. A generalized theory of angular momentum and atomic magnetism was soon developed that proved capable of accounting for the vast body of spectral data, including in particular the anomalous Zeeman effect.

Quantum Mechanical Perspectives

Here we need note only three aspects. First, the properties of a general angular momentum vector **J**, which can be orbital, spin, or a combination of both, differ substantially from the old quantum theory. As illustrated in figure 3, the pertinent features can be visualized in terms of a semiclassical model. As a consequence of the uncertainty principle, only the magnitude |**J**| of the angular momentum vector and its projection m_J on some axis (denoted Z) can be simultaneously specified (both in units of $h/2\pi$). The magnitude is given in terms of a quantum number J; depending on the nature of the system, it can take either half-integral or integral nonnegative values, i.e., $J = 1/2, 3/2, \ldots$ or $J = 0, 1, 2, \ldots$. For any given J, the projection m_J has $2J + 1$ allowed values, running in integer steps from $-J$ to $+J$ (and including zero if J is integral). Since an angular momentum state is fully defined by its values of J and m_J, an appropriate model has the **J** vector precessing uniformly about the Z axis, with the angle α between **J** and Z determined by J and m_J.

In contrast to figure 1, space quantization now refers to the $2J + 1$ allowed values of m_J or the angle α, and not to quasiplanetary orbits, which are banished in quantum mechanics by the wave properties of the electron as well as the uncertainty principle. In particular, the direction of **J** can never coincide with the Z axis (i.e., $\alpha \neq 0°$ or $180°$). Furthermore, unlike the old quantum theory (where $L = 0$ or $m = 0$ were considered meaningless), zero values of the orbital angular momentum and its projection are allowed. Space quantization now emerges as a universal prop-

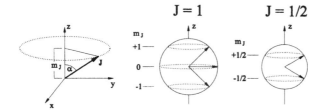

Figure 3: Space Quantization According to Modern Quantum Theory. At left is shown the semiclassical model for an angular momentum vector **J** (orbital, spin or resultant of both): the **J** vector precesses uniformly about a space-fixed axis Z (for example, a magnetic field), with projection m on that axis. In accord with the uncertainty principle, the azimuthal angle around Z is unobservable, and the magnitude $|J|$ of the angular momentum (in units of $h/2\pi$) is given by $[J(J+1)]^{1/2}$, with J the angular momentum quantum number. Also shown are the space quantizations for $J = 1$ and $J = \frac{1}{2}$, including the allowed values of the projection quantum number m, which run in integer steps from $-J$ to $+J$; thus there are a total of $2J + 1$ discrete orientations of the **J** vector.

erty associated with a disembodied angular momentum vector; for example, for $J = 0, 1, 2, 3, \ldots$ the number of discrete orientations are respectively: $1, 3, 5, 7, \ldots$; whereas for $J = 1/2, 3/2, 5/2, \ldots$ the number of orientations is $2, 4, 6, \ldots$.

Second, the apparent agreement of the Stern-Gerlach splitting with old quantum theory is now seen to be a lucky coincidence. For the silver atom, the orbital angular momentum of the valence electron is actually zero (not one unit as presumed in the Bohr model). The magnetic moment is due solely to one half unit of spin angular momentum. That produces the twofold splitting (with $m_s = \pm 1/2$, in contrast to the $m = \pm 1$ components of the Bohr model). Moreover, the spin produces a magnetic moment of the same size as one unit of orbital momentum would (that is, one Bohr magneton), by virtue of a factor of two arising from a relativistic effect not recognized until 1926. There was thus an uncanny cancellation of errors.

Also unwittingly lucky was the choice of silver, an atom with J a half-integer.[22] If the atom instead had J an integer, one of the space-quantized components would have a zero value of the projection quantum number, $m_j = 0$. That component would not be deflected; hence on the collection plate it would occupy the gap between the deflected components. Beam splitting thus would have been undetectable in the original Stern-Gerlach experiment, a result that would have appeared consistent with classical mechanics.

Third, quantum mechanics disposed of many vexing conceptual puzzles of the old quantum theory. Among these is magnetically induced birefingence, the issue that galvanized Stern. With electrons quivering in waves rather than moving in quasiplanetary orbits, there no longer exists the drastic asymmetry of the old model, illustrated in figure 1, which would have produced strong double refraction.[23] Another question rendered irrelevant is how the silver atoms take up definite spatial orientations in the field, which stymied Einstein and Ehrenfest. Quantum mechanics does not permit the orientation process to be visualized as twisting into position the atomic magnets, since the uncertainty principle prohibits following the change of state of an atom entering the field. The modern description deals only with the probability distribution of the atoms between the space-quantized states defined by the field.

There persists an intriguing historical puzzle, however. Since in 1922 the Stern-Gerlach splitting aroused much interest, prompting testimonials of its importance from leading physicists, we expected that the discovery of electron spin in 1925 would very soon have led to a reinterpretation of the splitting as due to spin. However, in a search of the contemporary literature, the earliest attribution of the splitting to spin that we have found did not appear until 1927.[24] Perhaps this is merely another instance of *sic transit gloria mundi*; but the hiatus seems surprising in view of the rapid flowering of quantum mechanics in those years. A host of current textbooks mention the Stern-Gerlach splitting as demonstrating electron spin. Of course, that is correct, but we have not found any text that points out the experimenters had no idea it was spin they had discovered.

Molecular Beams and Other Marvels

Late in 1922, Stern became professor of physical chemistry at Hamburg, where he undertook an ambitious program to develop further molecular beam methods. From this came many basic techniques and germinal results.[25] The crowning achievements were a quantitative confirmation of the wave nature of matter, by diffraction of a helium beam from a crystal, and the first measurements of nuclear magnetic moments, for the proton and deuteron. Those nuclear moments are smaller than that of an electron by a factor of roughly 1/1000, and so required a greatly improved magnetic deflection experiment. Beforehand, theorists advised that the experiment would be wasted effort, as the factor would surely just equal the ratio of the nuclear mass to the electron mass. Stern found a much different result, however, which revealed that the proton and deuteron were not elementary particles, but must have internal structure. In the summer of 1933, shortly after this epochal finding, the Nazi nightmare forced Stern to emigrate. He never regained a pacesetting role in research. Yet his work abides, and with it the inspiring example of his ardent pursuit of lucid understanding.

Descendants of the Stern-Gerlach experiment and its key concept of sorting quantum states via space quantization are legion. Among them are the prototypes for nuclear magnetic resonance (NMR), radioastronomy, atomic clocks, and the laser. A family tree for these and kindred developments has been traced by John Rigden in his biography of Isidor I. Rabi;[26] it has been examined also by Holton in his model for the growth of scientific research.[27] The tree sprouted in Stern's Hamburg laboratory, where as a postdoctoral visitor in 1928 Rabi became captivated by the molecular beam method, and it flourished when Rabi transplanted it to American soil.[28] The fabulous harvest, still being reaped, now includes ways to probe nuclei, proteins, and galaxies as well as the means to image bodies and brains; perform eye surgery; read music or data from compact discs; and scan bar codes, grocery packages, or DNA base pairs in the human genome. All these marvels and many more stem from exploiting radiative transitions between space quantized quantum states.

As both Stern and Rabi began in physical chemistry, our own field, we take note also of a hearty offshoot in which molecular beam techniques, augmented both by magnetic and electric resonance spectroscopy and by lasers, have evolved into powerful tools for the study of molecular structure and reactivity. Particularly welcome is the ability to examine individual reactive collisions, by colliding beams in a vacuum and detecting the products in free flight, before subsequent collisions degrade the information they carry about the intimate dynamics of the reaction. Such methods, applied and refined in many laboratories over more than three decades, have enabled the forces involved in making and breaking chemical bonds to be resolved and related to the electronic structure of reactant molecules.[29] In recent years, to enhance both collision and spectroscopic experiments, much work has been devoted to controlling the spatial orientation of molecules by means of external electric or magnetic fields.[30] Instead of an electron in an atom, these methods deal with space quantization of the molecular rotation, or end-over-end tumbling. Since this is analogous to orbital motion, and since the wave character is much less pronounced for a molecule than an electron, this recent work is actually quite closely linked to the concepts invoked seventy-five years ago by Stern.

The legacy of space quantization has been profound and pervasive in theoretical chemistry. The spatial distribution of electrons is crucial to the explanation of not only the periodic table of elements but also major aspects of chemical bonding and reactivity. As a benediction, we mention another historical link with Stern. In his 1913 papers, Bohr began a quest to explain chemical periodicities in terms of electronic structure; by 1922, he had attained some success (leading, for example, to the discovery of element 72, Hafnium, named for Copenhagen). This was largely based on the "shell structure" imposed by the available space-quantized states, as then understood. In late 1924, Pauli, who had been recruited by Stern as *privatdozent* for theoretical physics at Hamburg, made a decisive advance.[31] In effect, he invoked space quantization of both the orbital and spin angular momenta of electrons.[32] Pauli employed, in addition to the familiar three quantum numbers, a fourth that could take only two values (say, "up" or "down"). He added a key postulate, known since as the "Pauli exclusion principle": no two electrons in an atom can have the same value for any of the four quantum numbers. Accordingly, for any principal quantum number, not more than two electrons (one up, one down) can occupy each of 1, 3, 5, 7, ... distinct space quantized states.[33] This proved to be the final step needed to explicate the pattern of the periodic table, completing a long odyssey extending back over fifty years. Likewise, the Pauli principle also accounted for the prevalence of electron pairing in chemical bonds, a rule proposed heuristically in 1916 by Gilbert Newton Lewis. The exclusion principle, in a generalized form, is now recognized to stem from deep symmetry properties at the core of modern theoretical physics. But its discovery sprang from confronting chemical questions with space quantization.

ACKNOWLEDGMENTS

This paper is dedicated to Gerald Holton, to express our admiration of him as a humanistic scholar and dedicated teacher. His many insightful case histories, especially his studies of Einstein, have elucidated science as an intellectual adventure and cultural force, replete with thematic presuppositions, creative imagination, and varieties of rhetoric as well as crucial experiments.[34] Holton has also forthrightly addressed current nihilistic views of science, "not in the abstract, but in the natural setting of specific historical cases."[35] We hope that our account of the Stern-Gerlach saga may serve to complement some of Holton's work. In kindred fashion, it emphasizes how untidy and uncertain frontier science usually is—often hampered by misleading conceptions, yet capable of opening up new domains of understanding. The process is easily misconstrued, as recently seen in the so-called science wars.[36] In our view, most of the belligerence is unwarranted. As in science itself, current foolishness and errors will be subject to the scrutiny of a coming generation of scholars. May they chuckle rather than growl. Meanwhile, with Holton, we urge skeptics and advocates alike to ponder the lessons and legacy of episodes such as the Stern-Gerlach story, here offered in its "natural setting."

ENDNOTES

[1]Gerald Holton, *The Advancement of Science, and Its Burdens: The Jefferson Lecture and Other Essays* (Cambridge University Press, 1986), 197.

[2]Unless otherwise cited, quotations from Bohr or Einstein and literature references to papers mentioned can readily be found in one or another of three splendid books by Abraham Pais (all published by Oxford University Press): *"Subtle is the Lord...": The Science and the Life of Albert Einstein* (1982); *Inward Bound: Of Matter and Forces in the Physical World* (1986); or *Niels Bohr's Times* (1991). Curiously, Pais makes only glancing reference to the Stern-Gerlach experiment, in a couple of footnotes.

[3]Emilo Segre, "Otto Stern," in *Bibliographical Memoirs of the National Academy of Sciences* 43 (1973), 215–236. References for quotations and Stern's work not otherwise cited can be found in this article.

[4]K. Mendelssohn, *The World of Walther Nernst* (London: MacMillan Press, 1973), 95.

[5]F. Hund, *Geschichte der Quantentheorie* (Bibliographisches Institut, Mannheim, 1975). According to Hund, Pauli dubbed this vow the "Ütli Schwur," a nod to the legendary "Rütli Schwur" of Wilhelm Tell, which bound together some of the Swiss cantons. Yet von Laue, a Nobel laureate in 1912, was among the first (from 1919 on) and most persistent to nominate Bohr for the Nobel Prize, which Bohr received in 1922; Bohr did likewise for Stern, who received the prize in 1943. See Pais, *Niels Bohr's Times*, 213–216.

[6]Pais, *Niels Bohr's Times*, 146.

[7]For a pertinent analysis, see Arthur I. Miller, *Imagery in Scientific Thought: Creating 20th-Century Physics* (Boston: Birkhaüser Boston, 1984); cf. 132–133.

[8]Peter Galison, *How Experiments End* (Chicago: University of Chicago Press, 1987), 50.

[9]Wolfgang Pauli, "Remarks on the History of the Exclusion Principle," *Science* 103 (1946): 213.

[10]A volume commemorating the centennial of Stern's birth contains an English translation of his 1921 paper. See *Zeitschrift für Physik* D 10 (1988): 114–116.

[11]In 1960, Otto Stern was retired and living in Berkeley, where one of the authors (D. H.) had the opportunity to hear Stern reminisce about his career. This is not an actual quotation from Stern but is cast in a first-person, "as told to" form in an attempt to capture his way of telling stories. Fuller versions are given in Dudley Herschbach, "Molecular Dynamics of Elementary Chemical Reactions," *Angewandte Chemie International Edition in English* 26 (1987): 1225.

[12]J. Mehra and H. Rechenberg, *The Historical Development of Quantum Theory* (New York: Springer, 1982), 435.

[13]Ibid. The "very good mechanic" was Mr. A. Schmidt.

[14]Walther Gerlach, "Otto Stern zum Gedenken," *Physikalische Blätter 25* (1969): 412; "Zur entdeckung des 'Stern-Gerlach-Effektes,'" *Physikalische Blätter 25* (1969): 472.

[15]Stern, "as told to Herschbach" (see note 11), 1225.

[16]Gerlach, "Zur entdeckung . . . ," 473.

[17]W. Schütz, "Persönliche Erinnerung an die Entdeckung des Stern-Gerlach-Effektes," *Phys. Blätter 25* (1969): 343.

[18]Max Born, *My Life: Recollections of a Nobel Laureate*, (London: Taylor & Francis, 1978), 195.

[19]Gerlach also sent in early 1922 a photograph of the collector plate showing the beam splitting to Niels Bohr as a postcard, with the message: ". . . attached [is] the experimental proof of directional quantization. We congratulate [you] on the confirmation of your theory." Front and back views of the postcard are shown in A. P. French and E. F. Taylor, *An Introduction to Quantum Physics* (New York: Norton, 1978), 437.

[20]In the Stern centennial volume, Isidor Rabi (as told to John Rigden) recalls, "As a beginning graduate student back in 1923, I . . . hoped with ingenuity and inventiveness I could find ways to fit the atomic phenomena into some kind of mechanical system. . . . My hope to [do that] died when I read about the Stern-Gerlach experiment. . . . The results were astounding, although they were hinted at by quantum theory. The separation of the beam of silver atoms into two components occurred as if these moments pointed either one way or the opposite way. There was no mechanism that would orient them in one way or another since on leaving the source they were arranged quite statistically . . . the whole thing was a mystery. . . . This convinced me once and for all that an ingenious classical mechanism was out and that we had to face the fact that the quantum phenomena required a completely new orientation." See *Zeitschrift für Physik* D 10 (1988): 119.

[21]Since Stern was Einstein's "first pupil," it seems apt to note here that magnetism and spin are both consequences of relativity. For magnetism this was shown in 1912 (the year Stern joined Einstein) by Leigh Page, a young Yale professor; a lovely discussion of Page's paper is given by Edward M. Purcell, in H. Woolf, ed., *Some Strangeness in the Proportions: A Centennial Symposium to Celebrate the Achievements of Albert Einstein* (Reading, Mass: Addison-Wesley, 1980). The relativistic origin of spin was shown in 1928 by Paul Dirac; his awesome work is well described by Pais. For a delightfully attractive, nontechnical treatment of all things magnetic, see James D. Livingston, *Driving Force: The Natural Magic of Magnets* (Cambridge, Mass: Harvard University Press, 1996).

[22]Furthermore, while Stern selected silver because it had a single valence electron, in 1921 he could not be certain that its inner electrons form a "closed shell," i.e., are paired up with all the spins and orbital angular momentum projections adding up to zero.

[23]Magnetically induced birefringence in gases, known as the Voight or Cotton-Mouton effect, does in fact occur. It is a very weak, secondary effect (quite different in origin and magnitude from that implied by the Bohr model). See, for example, A. D. Buckingham, W. H. Prichard, and D. H. Whiffen, *Transactions of the Faraday Society* 63 (1967): 1057 and R. Cameron, et al, *Physics Letters* A 157 (1991): 125.

[24]Ronald G. J. Fraser, "The Effective Cross Section of the Oriented Hydrogen Atom," *Proceedings of the Royal Society* A114 (1927): 212. This paper summarizes experimental evidence that the ground state of several atoms, including hydrogen, sodium, and silver, are isotropic, contrary to the Bohr-Sommerfield model. These results agree with the 1926 wave mechanics of Schrödinger, according to which for these atoms the ground-state orbital angular momentum and associated magnetic moment are zero. Fraser concludes "that which orients" and thereby produces Stern-Gerlach splitting is "apparently" the spin magnetic moment.

[25]Norman F. Ramsey, "Molecular Beams: Our Legacy from Otto Stern," *Zeitschrift für Physik* D 10 (1988): 121.

[26]John S. Rigden, *Rabi: Scientist and Citizen* (New York: Basic Books, 1987). This superb biography has much material about Stern and the impact of the Stern-Gerlach experiment; see especially 46–65.

[27]Gerald Holton, *Thematic Origins of Scientific Thought: Kepler to Einstein,* rev. ed. (Cambridge, Mass.: Harvard University Press, 1988).

[28]Another American physicist luckily influenced by Stern was Ernest Lawrence. They met in 1929, on coincident visits to Harvard during Christmas time. Unaccustomed to Prohibition, Stern asked Lawrence to take him to a speakeasy. While contemplating the circular rings left by their wine glasses, Lawrence diagrammed an idea he had been mulling over for months, a means to accelerate ions in a magnetic field. Stern urged him to stop talking about it, get back to his lab at Berkeley, and work on the idea; Lawrence took the advice and soon developed his cyclotron. This story comes from interviews by Nuel Pharr Davis, *Lawrence & Oppenheimer* (New York: Simon & Schuster, 1968), 27–28.

[29]A recent survey, entirely nontechnical, is given by Dudley Herschbach, "The Shape of Molecular Collisions," in Martin Moskovits, ed., *Science and Society* (Concord, Ontario: House of Anansi Press, 1995), 11–28. The volume honors John C. Polanyi. Key references can be found in a recent *festschrift* issue (honoring Yuan T. Lee) of the *Journal of Physical Chemistry* A 101 (1997): 6339–6820.

[30]As yet, the only nontechnical account is by Bretislav Friedrich and Dudley Herschbach, "Spatial Orientation of Molecules," *Physics News* (1992): 14–15. References to research papers can be found in Hansjurgen Loesch, *Annual Review of Physical Chemistry* 46 (1995): 1147. Especially pertinent is recent work related to the Voigt effect; see Alkwin Slenczka, *Journal of Physical Chemistry* A 101 (1997): 7657.

[31]Mendelssohn, *The World of Walther Nernst,* 124.

[32]Although Pauli "in effect" was invoking electron spin (with his fourth quantum number specifying its two projections), nonetheless he adamantly rejected the idea that the electron could have an intrinsic angular momentum. He also dissuaded others, prior to Uhlenbeck and Goudsmit. This was doubly ironic. Despite his opposition to the notion of electron spin, he had been the first to suggest that nuclei might have spin. Also, although Pauli had a deep grasp of relativity, his error had to do with the relativistic description of the electron motion. See George F. Uhlenbeck, "Personal Reminiscences," *Physics Today* 29 (1976): 43.

[33]We are unable to resist noting an appealing coincidence. In Kyoto, Japan, the Ryoanji Temple has a garden facing the abbot's quarters which consists solely of four groups of rocks set on white sand: 1, 3, 5, 7; the garden dates from the Muromachi Period but its origin and any intended symbolism is not known. Matsuki Kokichi, ed., *The Gardens of Kyoto,* (Kyoto: Kyoto Shoin Co, 1987), 102.

[35]Gerald Holton, *Einstein, History, and Other Passions* (Reading, Mass.: Addison-Wesley, 1996).

[36] Gerald Holton, *Science and Anti-Science* (Cambridge, Mass.: Harvard University Press, 1993).

[37] For a sprightly survey and (in our opinion) sensible perspective, see Jay A. Labinger, "The Science Wars and the Future of the American Academic Profession," *Dædalus* 126 (4) (Fall 1997): 201–220.

Thus in the history of the sciences there is likely to be a period, their hinge, when they begin to come out of common sense, when they come to find that the common view of this experience is not an adequate explanation, when creative synthesis begins. That is the time when there is meat in scientific discovery to enrich human life. That is the time when the content of a science may indeed influence culture. Of course it is also the time of the abuses of scientific discovery of which we have heard so much.

—Robert Oppenheimer

"The Growth of Science and the
Structure of Culture,"
from *Dædalus* Winter 1958,
"Science and the Modern World View"

E. H. Gombrich

Eastern Inventions and Western Response

*So long as men believed that the Greeks and
Romans had attained, in the best days of their
civilization, to an intellectual plane which posterity
could never hope to reach, so long as the authority
of their thinkers was set up as unimpeachable, a
theory of degeneration held the field, which
excluded a theory of Progress.*

READ IN CONJUNCTION WITH THE TITLE I have chosen for this
essay, the above passage from J. B. Bury's classic study
The Idea of Progress[1] may suffice to indicate the hypothesis to be outlined: namely, that it was the Western response to
the technical inventions that had reached Europe from the East
that undermined and finally swept away the belief that, in Bury's
words, "excluded a theory of Progress."[2]

It was a momentous event in the history of Western thought
when the cyclical conception of history was replaced by the
image of one ascending line of human development. This reorientation cannot have been the result of a single cause, but it may
be that Bury and other historians of ideas have underrated the
role of Eastern inventions in this process.

The passage from Bury emphasizes that the movement we call
"The Renaissance" implicitly accepted the cyclical theory: The
so-called humanists strove to return to the pure Latin of authors

---

*E. H. Gombrich is Professor Emeritus (formerly Director and Professor of the History
of the Classical Tradition) at the Warburg Institute, University of London.*

such as Cicero, since they regarded the language and style of the classical age as the true manifestation of a civilized society.

In the very period, however, when the intolerance of the Ciceronians had become notorious, they were challenged by a grammarian from Arezzo: In 1471 Giovanni Tortelli published a book that at first sight may look to be of purely philological interest.[3] As a student of language he was concerned with Latin words deriving from the Greek, words such as *Horologium*, which means a clock. Remembering that the medieval clock, driven by springs, was unknown to the ancient world, he is tempted to launch upon a digression about the usefulness of this new invention. Once in his stride, he goes on to mention other novel inventions, such as the marine compass, not to speak of that terrible and miraculous weapon, the bombard, which he compares to a bolt of lightning; nor, he goes on, did the ancients have stirrups, or watermills, for which they had no words. The same applied to cotton, for which he could not find any agreed Latin term, and to that marvelous musical instrument the organ. His list continues rather unsystematically, referring to the introduction of sugar and of the candle. Neither did the ancients know falconry, nor the process of gilding, nor the art of the *niello*, a recent invention of the goldsmiths. Spectacles are new and so are blowguns. Silk was a rarity in antiquity.

All this goes to show, he concludes, that new words have to be found for new things, and the author is happy to refer to the authority of Priscianus, who had said that if the timidity of writers prevented their use of new words that had become necessary, the Latin language would forever be condemned to languish in a narrow prison. It would become what we call it now: a dead language.[4]

So here we have a Renaissance humanist who has discovered that the aspirations of his fellow humanists are impracticable: one cannot bring back the language and style of Cicero, because in our world one has to talk of things Cicero never dreamt of—and many of them, as we know, originated in the East.

Admittedly it took a number of generations after Tortelli for the realization to sink in that the classical past could never be recovered, and that a new age had dawned. It is perhaps fitting that the most eloquent testimony to this dramatic insight came

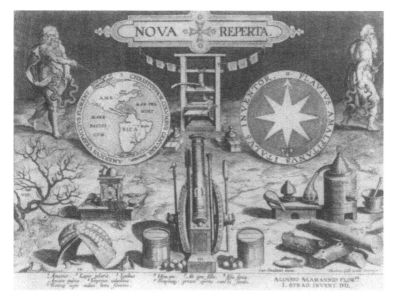

Figure 1: Jan Van der Straet (Johannes Stradanus), *Nova Reperta*, title page.

from Florence, the very city that considered itself the cradle of the Renaissance movement. I refer to the series of prints entitled *Nova Reperta* (New Inventions), dating from the end of the sixteenth century, designed by the Flemish expatriate Jan Van der Straet (1536–1605), who latinized his name to Johannes Stradanus.[5] Stradanus worked at the court of Francesco de Medici, who was not only a patron of the arts but an active promoter of science and industry, a fact that is not irrelevant to my subject.[6]

The series was apparently done in two installments, the first, of nine prints engraved by Theodoor Galle, and the second, of ten additional inventions, originally engraved by J. Collaert and at some later date by Philips Galle, one of whose captions refers to an invention only published in 1599.

Some of the individual prints in this series have been illustrated on and off in histories of technology, but my concern here is with the *Nova Reperta* as a document of the new interest in those inventions that distinguish the artist's own period from antiquity.[7] The title page (figure 1) is quite explicit: The words *Nova Reperta* appear within a cartouche displaying a symbolic rendering of the Southern Cross. Two allegorical figures hold

Figure 2: Jan Van der Straet (Johannes Stradanus), Mola Alata (Windmills), from *Nova Reperta*.

that ubiquitous symbol of a serpent biting its own tail (believed to have been an Egyptian hieroglyphic), which can only be interpreted as signifying Time.[8] The youthful figure who comes onto the scene points to a representation of the American continent, within a roundel of the globe marked "Christophor. Columbus Genuens. inventor. Americus Vespuccius Florent. retector et denominator"; we still reckon the modern age from the discovery of America.[9] On the other side, an old man with a long beard is leaving the stage. The old age is departing, a new age has arrived—an age marked by a series of new discoveries and inventions that are explained in the captions beneath, as a kind of trailer to the series: I. America; II. The mariner's compass (here wrongly attributed to one Flavio of Amalfi); III. A mounted cannon with cannon balls and barrels of gunpowder; IIII (IV). A printing press; V. A clock; VI. The wood and bark of a tree that he calls *hyacum* (an alleged remedy for syphilis); VII. A still; VIII. Silkworms and their cocoons on a mulberry tree; and IX. Stirrups, attached to a saddle. At a later date Stradanus added another nine, which are more of a mixed bag: windmills (figure

Figure 3: Jan Van der Straet (Johannes Stradanus), Saccharum (Sugar; also showing oil press), from *Nova Reperta*.

2); watermills, the olive press, and cane sugar (figure 3); oil painting, spectacles, the establishment of longitude, the polishing of armor, the astrolabe, and the art of engraving on copper (figure 4).

Though Stradanus's list is long and rather mixed, he missed two important inventions that he had in front of his nose: the paper on which his series was printed and the numerals that listed them. The invention of paper in China and the story of its adoption by the West belongs to the best-documented episodes in the history of technology. What we call "arabic" numerals are known to have come from India, and therefore can also be classed as an Eastern invention.

It is common knowledge that, for the first three of Stradanus's new inventions (the compass, gunpowder, and printing), the Chinese had priority over the West. Among the other innovations listed this also applies to silk, and possibly to watermills. The Greco-Roman world was far advanced in the construction of mechanical gears, which were one element in the mechanism of clocks, but they lacked the vital contribution of the escape-

Figure 4: Jan Van der Straet (Johannes Stradanus), Lapis Polaris Magnes (Magnetic
Compass, also showing other inventions), from *Nova Reperta*.

ment that secures even movement.[10] In his Wilkins lecture of
1958, Joseph Needham established beyond reasonable doubt
that the Chinese had priority over the West in the construction
of such a device, but it is possible that an analogous mechanism
was developed independently in the West in the thirteenth century. An Eastern (though not necessarily Chinese) element is
certain in the case of windmills, stirrups, and sugar, and an
Arabic influence is at least possible in the case of distilling and
the astrolabe. Inventions on Stradanus's list that certainly originated in the West include oil painting, spectacles, the polishing
of armor and the art of engraving on copper. In his list there are
two false hopes: *hyacum*, or *guiacum officinale,* as a cure for
syphilis, and the establishment of longitude.

The history of technology in which these questions are ventilated
has become a large and specialized field. It was pioneered by Joseph
Needham, whose monumental work *Science and Civilization in
China*[11] is still a mine of important information, although the
debate about priority continues—for which see Lynn White's *Medieval Religion and Technology*, a book that is equally indispensable

to the historian of technology.[12] Fortunately, the argument of this essay is not dependent on the exact attribution of competing claims. What matters more in my context is the fact that the most significant innovations were developed in the context of magical practices. The magnetic compass, for example, had originally something to do with geomancy. Gunpowder may have been used to frighten away evil demons, and even printing is known to have been used to multiply Buddhist prayers to make them more effective. A similar suggestion has also been made for the first wind-driven mechanisms, which may have begun in Tibet as prayer wheels before they were used for irrigation. Both in the East and in the West the search for a reliable clockwork is connected with the construction of armillary spheres, intended to match and predict the movement of the heavens, rather than for today's purpose of accurate time-keeping.

We cannot tell how far Stradanus was aware of the origins of the inventions he illustrated, though the fact that he attributes the marine compass to a certain Flavio of Amalfi speaks against it; nor had Tortelli hinted at any foreign origin. Polydore Vergil, in *De Rerum Inventoribus*, merely states that the origins of many are unknown to him:

> There be many other things, whose Authors for Antiquity cannot be known; and some, because of the negligence of men, that will not write such things. No man can tell who began Clocks, Bells, the Ship-man's compass, Gowns, Stirrops, Caps or Bonnets ... Water-Mills and Clavicymbals, Tallow-Candles, re-claiming of Hawks, Rings, with many others, which for the Anciety, or over-sight of men be in extream Oblivion.[13]

Louis Le Roy, in *De la vicissitude ou variété des choses en l'univers* (published in 1577), mentions that although the invention of printing is generally attributed to the Germans, the Portuguese—who sailed all over the world and were trading in the remote East—had brought back books printed in the language and writing of those countries and reported that printing had long been used in those parts.[14] But other writers were explicit in their attribution of a number of these inventions to sources outside Europe: In Campanella's utopia, *La Città del Sole*, published in 1603, the visitor is told that the city's inhabitants always sent out explorers over the whole earth to study the customs, forces, laws, and histories of all nations, bad and good alike, and he learns that guns and printing existed in China earlier than in Europe.[15] Finally, Samuel Purchas, writing in *The Pilgrim*, says:

> Others, therefore, looke further unto the East, whence the Light of the Sunne, and Arts, have seemed first to arise to our World; and will have Marco Polo the Venetian above three hundred yeeres since to have brought it out of Mangi (which we now call China) into Italy. True it is, that the most magnified Arts have there first been borne, Printing, Gunnes, and perhaps this also of the Compasse, which the Portugals at their first entry of the Indian Seas found amongst the More, together with Cards and Qudrants to observe both the Heavens and Earth.[16]

Francis Bacon had come to a similar conclusion from these suggestions and rumors in his *Novum Organum* of 1620:

> Consider the force and effect of inventions which are nowhere more conspicuous than in those three which were unknown to the ancients, Namely printing, gunpowder and the magnet. For these three have changed the appearance and condition of the whole world, the first in letters, the second in warfare and the last in navigation, and from these there sprang innumerable changes so that no empire, sect or star appears to have exercised a greater power and influence on human affairs than these mechanical matters.[17]

In his utopia, the *New Atlantis*, which was published posthumously in 1627, Bacon also describes the famous House of Solomon, where wise men apply new knowledge, and where he was told that the Institution had in its service twelve spies "that Sayle into Forraine Countreys . . . Who bring us the Bookes, and Abstracts, and Patternes of Experiments of all other Parts. These they call 'Merchants of Light.'"[18]

It is a beautiful term, anticipating the potent image of the Enlightenment, which was to dominate the idea of scientific progress in subsequent centuries. What Bacon's program here exemplifies is a tendency of the human mind of which he was well aware: It is the tendency to go beyond the evidence, to jump to conclusions—what he called *anticipatio mentis*, and we now call extrapolation from the known to the unknown. This tendency encourages the belief that, if new inventions (*nova reperta*) have transformed the age, more such inventions will transform it even more thoroughly and will bring about that perfect mastery over nature that Bacon desired—and actually thought to be just round the corner.

There is an anthropological parallel that lends support to this hypothesis; I am referring to the strange phenomenon known as "cargo cults."[19] These are waves of excitement known to seize so-called primitive societies upon their contact with Western merchants. The goods these people saw arriving appeared to carry the promise of miraculous changes—usually proclaimed by a self-appointed prophet who announced the imminence of a new era of plenty, when cargo would arrive in tremendous quantities and the downtrodden would inherit the earth from the white man.

The reader may find this comparison somewhat far-fetched, but it so happens that a similar wave of irrational hope can be documented from the decades after Stradanus and Bacon had predicted the dawning of a new age. In her interesting book *The Rosicrucian Enlightenment*, my former colleague at the Warburg Institute, Frances Yates, concentrated on pamphlets and prophecies that circulated in England in the early seventeenth century, purporting to emanate from a secret society that may never have existed: the Society of the Rosy Cross. These pamphlets proclaim, in solemn language, the imminence of a new dawn, and explicitly base the prophecy on the new discoveries:

The only wise and merciful God in these latter days hath poured out so richly his mercy and goodness to mankind, whereby we do attain more and more to the perfect knowlege of his son Jesus Christ and Nature, that justly we may boast of the happy time, wherein there is not only discovered unto us the half part of the world, which was heretofore unknown and hidden, but he hath also made manifest unto us many wonderful, and never heretofore seen, works and creatures of Nature, and moreover hath raised men, imbued with great wisdom, who might partly renew and reduce all arts (in this our age spotted and imperfect) to perfection.[20]

Similar millenarian hopes and prophecies can certainly be found in many periods, but what is significant is the reference to new discoveries and new knowledge that we associate with the idea of progress, the subject of Bury's book quoted at the outset of this essay. If the earliest illustration of this potent idea was Stradanus's series, perhaps the most telling was the imagery of the Great Seal of the United States of 1776, embodying the aspirations of the Founding Fathers with its caption "*Novus ordo Seclorum*"—still to be seen on every dollar bill.[21]

It has become the fashion to level the charge of Eurocentricity at the West for ignoring our debt to the achievements of other civilizations. Yet while fully acknowledging this debt, we must still ask why the West, after the end of the Middle Ages, so rapidly overtook the great civilizations of the East.

In the venerable civilizations of the East, custom was king and tradition the guiding principle. If change came it was all but imperceptible, for the laws of Heaven existed once and for all and were not to be questioned. That spirit of questioning, the systematic rejection of authority, was the one invention the East may have failed to develop. It originated in ancient Greece.[22] However often authority tried to smother this inconvenient element, its spark was glowing underground. It was that spark, perhaps, that was fanned into flame by the awareness that our ancestors did not have the monopoly of wisdom, and that we may learn to know more than they have if only we do not accept their word unquestioned. As the motto of the Royal Society (dating from 1663) has it, *Nullius in verba*—By nobody's word.[23]

In the prewar years, when the Warburg Institute was housed next door to the Imperial Institute of Science, I overheard two students at lunch. "How does he know it is a wave?" I venture to think that this kind of question was not often heard in ancient China or India. It only became possible thanks to the position of science in our culture.

ENDNOTES

[1]John Bagnell Bury, *The Idea of Progress: An Inquiry Into its Origin and Growth*, new ed. (New York: Dover Publications, 1955), 66.

[2]I first formulated this hypothesis in a lecture at the Warburg Institute in 1964, at a time when I greatly benefitted from the help and advice of my late colleagues Otto Kurz and Frances Yates. I subsequently presented it several times at various venues, which afforded me the opportunity of submitting it to some of the leading experts in the field. While Professor David Pingree's reaction was critical, he also acquainted me with the passage from Tortelli, which seemed to me to round off my argument. In view of my final conclusion, I was not surprised that the late Joseph Needham's reaction was rather negative. That of Lynn White, on the other hand, was more encouraging. I left the paper in a drawer, since I was much aware of my amateur status and the growing obsolescence of my bibliography. What finally decided me to overcome my hesitation was that the views here expressed also clash with prevailing intellectual fashions—fashions that deny or denigrate that element of Western thought that ultimately gave our culture the edge over the great civilizations of the East: the faith in progress.

[3]The reference is to the discussion of the term "horologium" in *De Orthographia*. See Alex Keller's "A Renaissance Humanist Looks at 'New' Inventions: The Article 'Horologium' in Giovanni Tortelli's *De orthographia*," *Technology and Culture* II (1970): 345–364.

[4]For the tenacity with which Latin resisted ossification and remained alive as a flexible tool for scientific discussions up to the eighteenth century, see the informative article by Tullio Gregory, "Pensiero medievale e modernità" in *Giornale critico della filosophia italiana* Anno LXXV (LXXVII), Fasc. II (May–August 1996): 149–173.

[5]See Jan van der Straet (Stradanus), *"New Discoveries (Nova Reperta)." The Sciences, Inventions and Discoveries of the Middle Ages and the Renaissance as Represented in 24 Engravings Issued in the Early 1580s by Stradanus*, ed. Bern Dibner (Norwalk, Conn.: Burndy Library, 1953).

[6]Stradanus, who was a pupil of Giorgio Vasari, actually contributed to the series of frescoes in Francesco de' Medici's studiolo in the Palazzo Vecchio, Florence (1577), where he represented *L'alchimia*. He appears in each of the two group portraits of Vasari and his assistants, in the *Salone dei cinquecento*. See Ugo Muccini, *Il salone dei cinquecento in Palazzo Vecchio*, (Florence: Le Lettere, 1990), 54, 126; also Piero Bargellini, *Scoperto di Palazzo Vecchio*, (Florence: Vallecchi, 1968). For his biography and his large output of prints see Georg Kaspar Nagler's *Allgemeines Künstler-Lexicon* (Leipzig: W. Engelmann, 1872–75); reprinted Vienna, 1924.

[7]For some of the drawings for this series, see Bensovich, "The Drawings of Stradanus . . . ," *Art Bulletin*, 1956, 249 ff.

[8]See my *Symbolic Images* (London: Phaidon, 1972), 158–159.

[9]"Discovered by Christopher Columbus of Genoa, rediscovered and named by Amerigo Vespucci of Florence."

[10]As testified to by the astonishing find of an astronomical or calendrical calculating device involving more than thirty gear-wheels in a shipwreck in the Mediterranean; see Appendix under "The mechanical clock."

[11]Joseph Needham, *Science and Civilization in China* [multivolume continuing work] (Cambridge: Cambridge University Press, 1954–), henceforth cited as "Needham."

[12]Lynn White, *Medieval Religion and Technology* (Berkeley and Los Angeles: University of California Press, 1986), 43–57.

[13]Quoted in Needham, vol. 1, 53.

[14]Louis le Roy, *De la vicissitude ou variété des choses en l'Univers* (1577, 2nd ed. 1584), quoted in Bury, *The Idea of Progress*, 44–45.

[15]Tommaso Campanella, *La Città del Sole* (1623), ed. and trans. Daniel L. Donno (Berkeley and Los Angeles: University of California Press, 1982), 37, 47.

[16]Samuel Purchas, *The Pilgrim*, pt. I, Bk. II, Ch. I (1), in *Hakluytus Posthumous, or Purchas his Pilgrimes, contayning a History of the World in Sea Voyages* (1625); quoted in Needham, vol. 4, part I, 245.

[17]Quoted in ibid., vol. I, 19.

[18]Francis Bacon, *New Atlantis: A Work Unfinished*, trans. William Rawley (London: J. F. and Sarah Griffin, 1664), 42.

[19]Ian C. Jarvie, *The Revolution in Anthropology* (New York: Humanities Press, 1964).

[20]Frances Yates, *The Rosicrucian Enlightenment* (London: Routledge & Kegan Paul, 1972), 238.

[21]See my paper "The Dream of Reason: The Symbolism of the French Revolution," in *British Journal for Eighteenth Century Studies* (Autumn 1979), quoting the *Encyclopedia Americana, XIII* (1957), 362; republished in *F. M. R.* no. 39 (Milan, 1989).

[22]See Karl R. Popper, *The Open Society and its Enemies* (London, 1946).

[23]Horace, *Epistles*, I.i.14: *Nullius addictus iurare in verba magistri / Quo me cumque rapit tempestas, deferor hospes* (Not bound to swear allegiance to any master / wherever the wind takes me, I travel as a visitor).

APPENDIX

A selective bibliography of notes on the origins of the inventions, in the order of their occurrence in the text.

The marine compass: Needham, vol. 4, 249–277; as used in geomancy, vol. 2, 361.

Gunpowder: James Riddick Partington, *History of Greek Fire and Gunpowder*, (Cambridge: W. Heffer and Sons, 1960).

Printing (including its use for the multiplication of Buddhist prayers): Thomas Francis Carter, *The Invention of Printing in China and its Spread Westward* (New York: Columbia University Press, 1925), 5 (henceforth cited as

"Carter"); also Paul Pelliot, "Les plus anciens Monuments de l'Ecriture Arabe en Chine," in *Journal Asiatique* (1913, IIe sér.): 17, 139.

The mechanical clock: Derek de Solla Price, *Gears from the Greeks* (Philadelphia: American Philosophical Society, 1974); Joseph Needham, Wang Ling, and Derek de Solla Price, *Heavenly Clockwork: The Great Astronomical Clocks of Medieval China* (Cambridge: Antiquarian Horological Society, 1960); and David S. Landes, *Revolution in Time: Clocks and the Making of the Modern World* (Cambridge, Mass.: Belknap Press/Harvard University Press, 1983).

"Hyacum": Owsei Temkin, "Therapeutic Trends and the Treatment of Syphilis before 1900," *Bulletin of the History of Medicine* XXIX (4) (1955).

Longitude: William J. H. Andrewes, ed., *The Quest for Longitude: The Proceedings of the Longitude Symposium, Harvard University, Nov. 4–6, 1993* (Cambridge, Mass.: Collection of Historical Scientific Instruments, Harvard University, 1996).

Distillation: R. J. Forbes, *Short History of the Art of Distillation* (Leiden: Brill, 1948); also Julius Ruska, "Ein neuer Beitrag zur Geschichte des Alkohols," *Islam* IV (3) (1913) and H. Diels, *Die Entdeckung des Alkohols* (Berlin, 1913).

Silk: Procopius, *History of the Wars*, ed. H. B. Dewing, vol. V, Loeb Classical Library (1914–40), 229–231; also E. Day, *Ars Orientalis* I (1954): 233.

The stirrup: A. D. H. Bivar, "The Stirrup and its Origins," *Oriental Art*, new series 1 (2) (1955): 61–65.

The watermill: Needham, vol. 1, 232.

The windmill: Needham, vol. 1, 245; also White, *Medieval Religion and Technology*, 79 and (on wind-driven prayer wheels) 47.

Sugar: Carter, 130.

Spectacles: Guido Pancirolli, *Rerum memorabilium libri II* (Antwerp: 1599, pub. 1612); also Needham, vol. 4, 20.

The astrolabe: Needham, vol. 4, 7; also Arno Borst, *Astrolab und Klosterreform an der Jahrtausendwende* (Heidelberg: 1989).

Paper: Carter, 5; also Charles Singer, ed., *History of Technology*, 5 vols. (Oxford: Clarendon Press, 1954–58).

"Arabic" numerals: Bibhutibhasan Datta and Avandesh Narayan Singh, *History of Hindu Mathematics, A Source Book* (Lahore: Motilal Banarsi Das, 1935); D. E. Smith, *History of Mathematics*, and F. Nao, *Notes d'Astronomie Syrienne*, quoted in Needham, vol. 1, 220; also Carter, 191.

The main point is that the human race has not yet found how to use its mind. We are getting at this realization through the sciences, but the sciences have as yet by no means furnished all the answers. One reason is that for the particular purposes of science an incomplete view is adequate, particularly because the sciences are comparatively so simple. But for the wider purpose of the humanities—the complete human scene in all its scope—some more drastic reconstruction is necessary. It is, for example, obvious that the involvement of the humanities with the whole verbal machinery of thought is much more intimate than that of the sciences. I would place as the most important mark of an adequately educated man a realization that the tools of human thinking are not yet understood, and that they impose limitations of which we are not yet fully aware. As a corollary it follows that the most important intellectual task for the future is to acquire an understanding of the tools, and so to modify our outlook and ideals as to take account of their limitations.

—P. W. Bridgman

"Quo Vadis?"
from *Dædalus* Winter 1958,
"Science and the Modern World View"

James S. Ackerman

Leonardo da Vinci: Art in Science

MY OBSERVATIONS ON THE INTERACTION of science and art in the work of Leonardo da Vinci are of a different kind from the more general themes being addressed by the other contributors to this volume. However, the grandiose sweep of Leonardo's efforts to establish new scientific procedures may raise them to a comparable level, making them relevant to a study of the borderlines between science and other aspects of culture.

Leonardo made the faculty of vision—or more precisely, the gift and patience for intensive observation—the foundation of both his scientific investigations and his work as a figural artist. He was a protoscientist in the modern sense of what constitutes science, bringing to his investigation of the natural world not only an extraordinary artistic imagination, which led him to innumerable original discoveries, but a unique and idiosyncratic intellectual position that helped him to circumvent the mental blocks of his contemporaries.

Science in the century preceding Leonardo was based almost entirely on texts surviving from antiquity; experiment and the pursuit of new challenges was rare. The Aristotelian scientific tradition had been sustained by scholastic writers, primarily within the church. Humanist scholars, a new class comprised of teachers, poets, and court secretaries, sought to rediscover and edit Greek, Latin, and ultimately Hebrew texts and to improve literary style in these languages; while their primary interests were literary and historical, they also made available—often as

James S. Ackerman is Arthur Kingsley Porter Professor of Fine Arts Emeritus at Harvard University.

editors for the new printing houses—what had remained of mathematical and scientific treatises and their epitomes, such as those of Euclid and Archimedes, Galen and Ptolemy. They restored to circulation, in Latin, medieval Arabic texts; in the discipline of optics alone, this included the work of Avicenna and ibn al-Haytham, as well as their later Western heirs, Bacon, Vitellius and Pelacani.

Trained as a painter, sculptor, and designer of machines, Leonardo da Vinci was no humanist. At the start of his career he was unable to read the texts upon which he would have to base his scientific knowledge. He admitted that he had the reputation of an *omo sanza lettere* (an illiterate), meaning that he did not have a good command of Latin. During the 1490s in Milan, he struggled to improve his Latin; both as a result of this effort and of an increasing number of Italian epitomes, he acquired as much information as he needed in the innumerable fields of his interest.

Science in Leonardo's time was predominantly descriptive. The fields in which progress was made were those that could be investigated with the eye—anatomy, botany, cartography, zoology, and ornithology. Copernicus stands virtually alone in the two centuries prior to Galileo and Kepler in being productively engaged in theoretical science.

An astonishing number of studies and notebooks, only some of which have survived, records Leonardo's intense drive for a comprehensive knowledge of creation on the model of Aristotle. Like Aristotle, Leonardo was an empiricist, in contrast to adherents of the Platonic tradition who worked with logic and mathematics on abstract hypotheses conceived intellectually.

Leonardo starts from books, but in almost every field of investigation he moves from traditional explanation to one based on his own experiments and experience. Since early writing often copied from traditional texts, one cannot always be certain whether Leonardo himself agreed with a statement that he wrote down.

A vivid early drawing of a skull sectioned vertically and horizontally illustrates the point (figure 1). Although astonishingly precise in detail—there are a number of anatomical features overlooked in the earlier literature—major aspects of the draw-

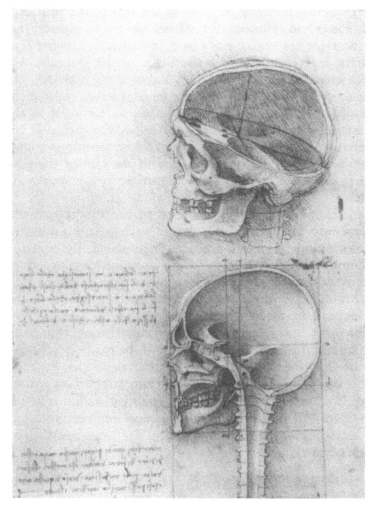

Figure 1: Leonardo da Vinci, Human Skull, Cut Horizonally and Vertically. Royal Library, Windsor, no. 19057. Courtesy of Her Majesty the Queen.

ing illustrate points determined theoretically rather than empirically. The grid of lines, for example, is intended to illustrate the conformity of the head to a system of privileged geometrical proportions. Further, the drawing illustrates the medieval doctrine that the vertical and horizontal axis of the skull must cross at the site of the *sensus communis*, or common sense, where all perceptions—of sight, sound, touch, and so on—were believed to be gathered; this was considered the seat of the soul. According to Plato and Hippocrates, whom Leonardo quotes, the soul must activate the entire body and, in particular, must transmit seeds for reproduction from the brain to the genitals. Accordingly, a channel must be provided through the marrow. The top of the spine in this drawing has a large interior channel, which Leonardo would not have found in his skeleton. If in this instance books triumphed over observation, the radically innovative character of the representation, which employs techniques of foreshortening just being devised at the end of the fifteenth century, is manifested by comparison to other works of this period.

In his early work, Leonardo tried to coordinate natural phenomena into an overarching design revealed in similarities of behavior among disparate natural phenomena. By contrast, experiments and observations in the later manuscripts focus on the diversity of nature: every form becomes determined by its function. An understanding of the extent to which Leonardo's scientific method matured over the course of his life has become possible only in the last twenty-five years through the painstaking determination of the chronology of the many notebooks and separate drawings.

For example, in an early sheet from the Codex Atlanticus (figure 2), Leonardo attempts to establish a unitary—percussion—theory of the transmission of sense impressions. The figures, drawn as similarly as possible, are accompanied by explanatory notes:

> How lights or rather luminous rays, can only pass through diaphanous bodies.
>
> How the surface x, o illuminated through point p, generates a pyramid that finishes in point c, and ends up at an other surface at r, s, which receives what is in x, o upside down. [This illustrates the *camera obscura*, a device discussed already by ibn al-Haytham in his eleventh-century treatise on optics.]
>
> If you put a colored piece in front of each light you will see the surface colored by it. [While this observation seems obvious to us, it served to demonstrate that light "rays" emanate from the object of vision and are not emitted from the eye, as some ancient writers had claimed.]

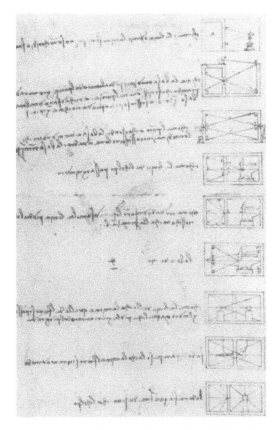

Figure 2: Leonardo da Vinci, The Percussive Action of Sense Impressions, Milan, Bibl. Ambrosiana, Codex Atlanticus, Fol. 126 *ra.*

How the lines of a blow pass through any wall. [Thus sound "rays" behave like their visual counterparts, except that they can pass through opaque barriers.]

How, finding a hole, many lines [of sound] spread; all others are weaker than *a-b.*

Voice in echo. [angle of incidence = angle of refraction]

How the lines of the magnet on iron are drawn in the same way.

Smell spreads the same way as a blow.

Every point generates infinite bases.

Every base generates infinite points.

Alternatively, Leonardo discovered such universal analogies in the behavior of moving matter. His almost obsessive effort to

Figure 3: Leonardo da Vinci, Study of the Flow of Water. Royal Library, Windsor,
no. 12660*v*. Courtesy of Her Majesty the Queen.

understand the action of water in response to a variety of constraints is the concern of a hydraulic engineer (figure 3). Indeed,
he was hired in that capacity in Milan and in Florence, where he
was charged to study the feasibility of making the Arno navi-
gable from Florence to the sea. His observations are concentrated primarily in two notebooks, Manuscript A of the Institut

de France, and the Leicester Codex, now the property of Bill Gates.

Probably the painstaking care taken in observing the movement of water throughout his career laid the groundwork for the late series of drawings depicting an earthly chaos, an Armageddon, in which the exploding landscape takes on equivalent forms—in this case, for art's sake. Like the transmission of sense impulses, the flow of liquid is also part of a universal scheme in early notes, as in this passage: "If a man has a lake of blood in him whereby the lungs expand and contract in breathing, the earth's body has its oceanic sea which likewise expands and contracts every six hours as the earth breathes." Leonardo goes on to associate underground springs with veins; a drawing in the Leicester Codex illustrates this.

These are the type of observations that support Foucault's characterization of the protoscience of Leonardo's time as based on similarities and analogies. Leonardo makes great claims for experiment, experience, and observation to distance himself from the scholastics and humanists who commented chiefly on texts, but in reality he was strongly directed by the textual tradition, constantly seeking—and only rarely finding—formulations of the causes of the effects he observed.

We may try to define the method with examples from two areas: first, physiological optics and visual perception; and second, anatomy, which Leonardo approached as a scientist, and only peripherally to help his art.

I begin with Leonardo's observations on the most widely used method of constructing painter's perspective in his time— *perspectiva artificialis*. The sketches reproduced in figures 4 and 5 illustrate his observation that when one gets close to the objects to be depicted, the rule by which objects appear to diminish in size as they are more distant from the eye no longer holds. The central ball or column in these drawings will be projected on the picture plane as smaller than those farther away, due to distortions resulting from a viewing point too close to the plane. This is an issue of perception; Leonardo's passion for observation gave him the capacity to challenge the dicta of artificial perspective, which were strictly geometric and abstract, unrelated to perception. They posited that all light rays converge

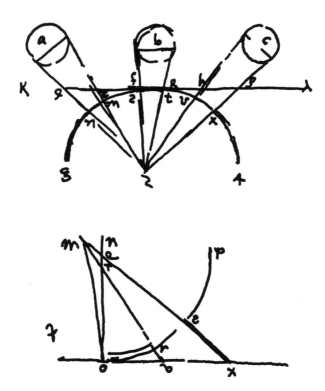

Figure 4: Leonardo da Vinci, Perspective Distortion of Nearby Objects. (Redrawn from a sketch in Ms. A of the Institut de France.)

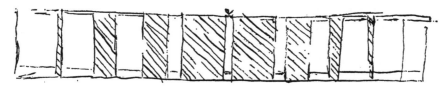

Figure 5: Leonardo da Vinci, Perspective Distortion of a Row of Columns Seen Close Up. (Redrawn from a sketch in Ms. A of the Institut de France.)

in a point at a hypothetical eye—a single eye in a fixed position, as Leonardo's diagram shows.

His answer was not to seek an improved construction procedure but to try to understand how the eye actually works. Over

Figure 6: Leonardo da Vinci, Diagram of the Eye's Reception of Images (above) compared
to the *camera obscura* (below). Institut de France, Ms. D, folio 8.

the course of the following decades, but mostly around 1505 and
at the end of his life, he tried to move beyond the literature of
medieval optics from ibn al-Haytham on, with which he was
thoroughly familiar, by devising experiments and trying to interpret their results. Two lines of investigation led him to conclusions that further diminished the hold of artificial perspective.
The first was derived from the use of the *camera obscura* as a
model of the eye (figure 6). The *camera* appeared in the early
drawing illustrating the propagation of sense stimuli (figure 2)
but was used there only to study the behavior of light rays. Later

Leonardo saw that the eye, with its lens and pupil, must function in the same way.

In figure 6, the two are compared on the same page from a small notebook—Manuscript D of the Institut de France—on optics and the physiology of vision. At the top of the margin Leonardo drew a horizontal section of the eye, and below, a horizontal section of a *camera obscura*, indicating that the two function in the same way. His notes on this page read:

> The experience [experiment] which shows how objects transmit their species or similitudes through an intersection inside the eye in the albugineous humor is demonstrated when species of illuminated objects penetrate through some small round hole [in an iron plate] into a very dark habitation. Then you will receive these images on a sheet of white paper placed inside this habitation somewhat near to this small hole, and you will see all of the mentioned objects with their true figure and colors, but they will be smaller and they will be upside down because of the said intersection (the paper should be very thin and seen from behind).

Most of Leonardo's drawings cut through the eye horizontally and therefore do not show that rays entering from above and below the aperture would also cross, casting an image upside down as well as reversed. The *camera*, apart from helping Leonardo to understand the physiology of vision, later became a tool for artists, who could sketch images received on the piece of paper.

In the section of the eye at the top of the manuscript page, we see Leonardo struggling with a problem he never solved; while the *camera obscura* reverses the image, we do not perceive the world as reversed. The answer, he thought, must be that some mechanism within the eyeball sets the image right by re-reversing it; in these drawings, it is the crystalline sphere, which is posited as a spherical lens. It is not the right answer; but what is right about Leonardo's understanding of vision is that it reveals the problematical nature of the theory of linear rays meeting at a point. This would lead to constructing pictures in quite a different way.

The rays do not come to a point because light rays are sensed on the whole surface of the cornea, but neverthe-

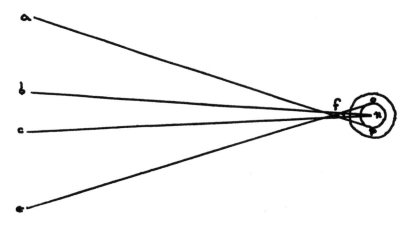

Figure 7: Leonardo da Vinci, Demonstration of the Capacity of the Eye to "See Through" a Small Nearby Object. (Redrawn from the Codex Atlanticus 250v.)

less pass through the pupil by being refracted. Leonardo's conviction that the entire cornea is sentient was based on an experiment demonstrating that a small obstacle placed directly in front of the pupil does not block a full view of the image before us. His explanation of this result is given in a sketch (figure 7). The top of the cornea will see it at position *e* and the bottom at *a*. The refraction, Leonardo reasoned, weakened the peripheral rays so that images looked fuzzier at the edges than at the center. But, as in medieval optics and in perspective theory, the central ray always has the sharpest and strongest effect because it hits the target unbent—as if, Leonardo says elsewhere, a bullet were to be shot *into* the barrel of a gun. If these studies did not lead directly to increasing modulation and haziness in Leonardo's painting and drawing, they must at least have urged him away from the sharp edges and strong local color of his fifteenth-century predecessors.

Let us now turn from optics to anatomy, and specifically to the evolution of Leonardo's investigations in the years following the drawing of the skull. The large size and striking conception of the drawing of the female torso

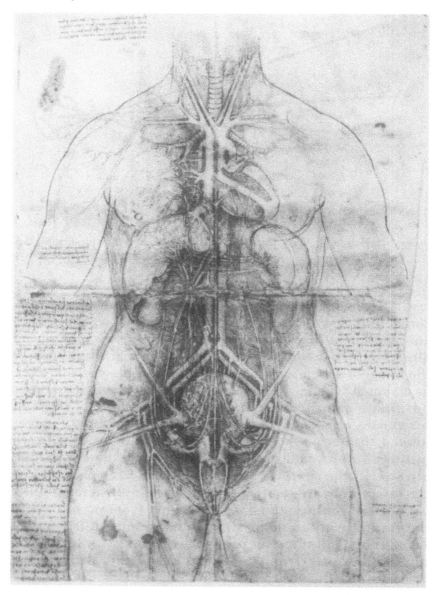

Figure 8: Leonardo da Vinci, Female Torso with Internal Organs. Royal Library, Windsor, no. 12291r. Courtesy of Her Majesty the Queen.

rendered transparent in order to reveal the viscera (figure 8) have made it the best known of Leonardo's second Milanese period in the years following 1506–08. It belongs to the large category of hypothetical studies combining animal dissection with traditional anatomy, making it a direct descendant of the medieval *situs* figure, of which two woodcuts from the 1495 printing of the anatomical work of Mundinus, showing the internal organs in a largely symbolic form, were the most recent examples. But while Leonardo's drawing appears to represent a great advance in accuracy over its predecessors, there is no sign of new empirical knowledge despite the fact that he had already performed human dissections; the progress is exclusively in drawing technique, particularly in the use of wash to enhance relief and in the combination of section, relief, and full and partial transparency. The uterus has "horns" to illustrate Mundinus's hypothesis that it is bound to the hips by two pairs of ligaments (one pair of which is a misunderstood transformation of the Fallopian tubes, which enter near the cervical region—at the bottom rather than at the top).

The celebrated image of the uterus, out of which a wedge has been cut to reveal a fetus within that looks like a year-old child, is one of a number of embryological studies dating from either 1510–13 or 1515, which places them among Leonardo's last anatomical observations. While the fetal position is accurately rendered, the drawing represents in other ways a reversion to those done before the turn of the century, based largely on speculation, with the addition of information gained from the dissection of a cow (whence the representation of the cotyledons binding the placenta to the uterine wall). Apparently Leonardo had not had the opportunity either to perform or to attend a Caesarian operation, though he does imply on a sheet from the same series that he had examined the fetus that appears in the drawings.

The artist's approach is entirely different in one of the last studies, done on a textured bluish paper, representing the heart

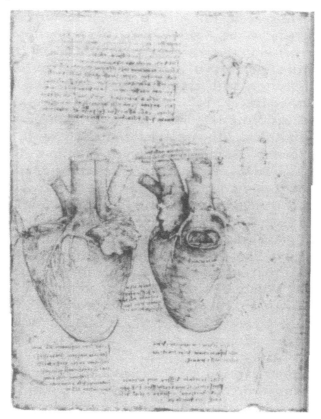

Figure 9: Leonardo da Vinci, The Heart of an Ox. Royal Library, Windsor, no. 17073. Courtesy of Her Majesty the Queen.

of an ox (figure 9). Here he was able to work entirely from direct experience. While speculation and traditional theory had a role in his understanding of pulmonary function, they did not impinge on his powers of observation, and in this sheet he made major advances applicable to the understanding of the heart. The drawings, showing the organ from the front and rear, exemplifies Leonardo's precept that the body parts should be represented in more than one view. It is the first anatomical description of the coronary arteries; to fully depict their form, Leonardo removed the pulmonary artery, and this served also to reveal the three cusps of the pulmonary valve.

The problem that most concerned Leonardo in investigating pulmonary function was how to explain the warming of the blood. In his early notes he suggested that it resulted from external forces of nature, comparing it to the rising of sap in trees. After 1508, he compared the heart to a stove that heats and then propels the blood through the veins. The heating process is explained in a note on the upper left of the drawing of the ox heart:

> The blood is more subtilized where it is more beaten, and this percussion is made by the flux and reflux of the blood generated from the two intrinsic ventricles of the heart to the two extrinsic ventricles called auricles ... which are dilated and receive into themselves blood driven from the intrinsic ventricles; and then they contract, returning the blood to those intrinsic ventricles.

The rendition of the heart is impressive not just as a record of significant advances in empirical observation—which, had they been known to others, might have led to an early discovery of the circulation of the blood—but for the extraordinary draftsmanship that reveals the object in a natural ambience of light and atmosphere. The very viscousness of the flesh is rendered in a way that has never since been matched in anatomical illustration.

For over twenty-five years of anatomical investigation, from the earliest representations of the skull to the late studies of the heart, Leonardo produced indelibly memorable images. But I should describe the skulls as marvels of draftsmanship—unsurpassed as such, yet of a much lower ambition. The drawings seek to give objects represented the palpability of sculpture and architecture, to make the pen virtually replace the skull itself. But conceptually they belong to the previous generation, with their concern for accommodating the object to external geometrical rules of perspective and proportion. Furthermore skulls, human though they may be, are inanimate objects, not different in essence from spheres with penetrations; in fact, they communicate nothing of the function of the human body. The heart, by contrast, is a lately pulsating organ of flesh and moisture, and not subject to the disciplines of mathematics. Such representations drew Leonardo far from the atmosphere of calculation and rationality of his early years into the almost indescribable mystery and complexity of life itself. One might ask why the depiction of the fetus, drawn only a few years before, looks more like the skull than like the heart. The answer is that it was an almost entirely hypothetical construction of the mind, on which the senses could be brought only peripherally into play; to draw it Leonardo had to fall back on the mode of theoretical construction. In fact, his experiments had taught him almost nothing of the process of gestation.

Whatever the progress made in the disciplines of physiology and anatomy following these drawings, it was not rivaled by progress in illustration. Anatomists after Leonardo and Vesalius were simply anatomists with, at best, rudimentary skills for recording their discoveries in images, and artists did not dissect people and oxen except perhaps to study musculature. In the five centuries since Leonardo performed his first dissection, the skill, artistry, and didactic potential of anatomical illustration has regressed, while the degree of advance in scientific accuracy has been in some respects less than Leonardo's advance over his immediate predecessors and followers prior to Vesalius.

But my point is that Leonardo was not simply an artist skilled in achieving verisimilitude; in this he had many rivals who, for all their talents, did not contribute significantly to anatomy. It is that his unbounded curiosity led him to pursue a vast range of natural effects and physiological responses, so he could bring to the recording of an animal heart an understanding of light, atmosphere, texture, and vision, as well as hydraulics for the flow of blood, botany for the branching of veins and arteries, mechanics for the expansion and contraction of the organ, and so forth. It was not simply Leonardo's grasp of these many natural processes that gives any one of his images a unique persuasiveness, but rather his desire to see all objects of his attention as manifestations of an overarching scheme. Yet "scheme" is not the right word, because it suggests something static, nor does the cliché "clockwork of the universe" fit, because it suggests something regular and mechanical. Leonardo's vision was of a breathing, mobile Nature—characterized by "flux and reflux" in the passage cited above, and revealed in the diurnal tides and, in microcosm, in the human body.

Do Leonardo's universal interests make him the fabled Renaissance man? Yes, in the sense of the range of his investigations, though not in the sense of his having achieved the kind of scholarly command of ancient texts that constituted the foundation of true knowledge for Renaissance humanists. In his scientific studies, Leonardo made earnest efforts to master the basics of traditional wisdom in each field; that meant knowing, at least indirectly, major Greek and Latin texts or modern summaries of them and, where relevant (as in the study of optics), medieval Arabic and Western works. Leonardo did not seek—as did his Florentine predecessor Leon Battista Alberti—celestial harmonies that might be emulated on earth; instead he sought a universal vital spirit animating all of creation. The two approaches may seem similar, but in fact they differ fundamentally. Alberti's is hierarchical, suggesting that we on earth try to emulate a celestial order in mathematical forms. Leonardo's, by contrast, is egalitarian. He posits that heaven, earth, man, and beast share and contribute to a mutually sustaining energy. It is more pagan than Christian.

Simultaneously with his achievements in describing objects in the natural world, Leonardo was opening new horizons in conveying the experience of vision itself. A small sketch of a copse of trees in the corner of another sheet at Windsor (figure 10) is a token of one of the most consequential changes in the history of Western art. Medieval and fifteenth-century drawn and painted trees, like those of Fra Angelico, Botticelli, or Leonardo's own early *Annunciation*, are discrete solid objects that one can count and distinguish from neighboring trees; they are as concrete as Leonardo's skull and come from a pictorial tradition that isolates every figure by its outline and local color. Leonardo approached the copse optically; he tried to catch the visual continuum at a particular time of day, as Monet would do four hundred years later. The trees are not individuals but the common recipients of a particular light and atmosphere. In a sense, it is inadequate to call the tree drawing "optical": every representation of nature could be called optical. But I mean to contrast optical to *conceptual* representation, which shows an ob-

Figure 10: Leonardo da Vinci, A Copse of Trees (red chalk). Royal Library, Windsor, no. 12431r. Courtesy of Her Majesty the Queen.

ject as one believes that one knows it to be, not as it appears to an interpreting personality at a particular moment.

It is tempting to say that figure 9 is objective in recording the heart as it really looks and that the copse is subjective in conveying a differentiated continuum of light and shade as experienced by an individual observer in particular temporal and physical conditions. But the image of the heart is also informed by those particularities, and one's visual and psychological faculties do not shift at will from an objective to a subjective mode of reception. I would rather suggest that the two are more alike than different in revealing the willingness of the artist to replace a conceptual approach to the world with an experiential one— leading to the end of a new art in the one sheet, and the end of a new science in the other.

No subject has engaged Dædalus *so frequently in its first forty years as education at all levels—elementary, secondary, and university. It is particularly appropriate that this anniversary issue should include an essay on American education. [S.R.G.]*

Patricia Albjerg Graham

Educational Dilemmas for Americans

PERIODICALLY, AMERICANS PAY ATTENTION to our children's educations.[1] Whenever we do so, inevitably we find ourselves deeply mired in contradictory accounts of their experiences. Newspapers abound with graphs revealing how poorly or how well US children take academic tests compared to each other and compared to others around the world. People seeking a simple indicator of a complex situation find a test score, and its relative standing among others tested, a handy guide of whether children are doing well or badly. Others, who are convinced that the public schools need fixing, seek a single solution—vouchers, perhaps, or charter schools—as the remedy. Still others observe that in some states there is five times as much spending on students in some school districts as in others ($15,744 versus $2,932 during 1994–95 in Illinois, for example) and argue that equalization of expenditure is the cure-all.[2]

Patricia Albjerg Graham is Charles Warren Professor of the History of Education at the Harvard Graduate School of Education and President of The Spencer Foundation.

Education is an important issue for nearly all Americans, but for almost none is it the most central issue in our lives. As an important but rarely preeminent question, we often discuss it superficially and seek one comprehensive solution. Our disappointment increases when no panacea materializes; but by that time, our sons and daughters are already grown and our interest in education has receded, since only a few of us worry about the educations of other people's children.

Why, then, does education present itself as such a persistent dilemma in the United States? I believe that there are three fundamental explanations: We change our minds about what the central tasks of schools should be; we want teenagers to get high school diplomas, but we are deeply ambivalent about what the content of their adolescent experience should be; and we are unsure how important school itself is in children's education.

First, we change our minds about what the central task of schools should be, and we expect schools to accommodate immediately to these shifting priorities. To a remarkable degree American schools historically have faithfully delivered what American society sought from them. When changes in society's expectations for the schools occurred, as has happened during the last dozen years, the schools have had difficulty making a complete and rapid adjustment to the new expectations. While American schools have always had a core commitment to academic achievement for some apparently gifted children, including children of the poor, the emphasis for the vast majority of the other students has changed significantly.

Schools do deliver what society wants, but slowly and incompletely. During this century society has set four principal but different goals for our schools:[3]

1900–25: *Assimilation.* Schools were the principal institution in which the many European immigrants and their children encountered an emerging and distinctive American culture. The schools understood that their primary mission was to "Americanize" the children into loyal and acculturated American citizens.

1925–54: *Adjustment.* The new and dominant progressive education movement, committed to the "whole child" (with special attention to the child's mental and social health), became codified after World War II as the Life Adjustment Movement in which 20 percent of students were to be educated for college, 20 percent for vocational training, and the remaining 60 percent with "general life skills," or how to adjust to life.

1954–83: *Access.* Again, after the 1954 desegregation decision in Brown v. Board of Education, schools were used as institutions to serve the broad goals of the society by providing access to those previously denied it. This emphasis continued with special attention to children of poor families in the Elementary and Secondary Education Act of 1965 and the Education for All Handicapped Act (PL 94-142) in 1975.[4]

1983–present: *Achievement.* The current effort to universalize academic achievement as a goal of the many and not just of the few is the most radical goal of all. The enunciation of this sentiment came most poignantly to the American people in the 1983 report of a Reagan-appointed commission on American education, entitled *A Nation at Risk.* Yet even in this last decade the increases in school budgets have been concentrated in special education, not regular academic education.

Richard Rothstein and Karen Hawley Miles recently examined expenditures for nine school districts between 1967 and 1991, which revealed that real school spending increased by 61 percent in that period. They found that the share of expenditures going to regular education dropped from 80 percent to 59 percent, while the share going to special education climbed from 4 percent to 17 percent. Of the net new funds spent on education in 1991, only 26 percent went to improve regular education, while about 38 percent went to special education for severely handicapped and learning-disabled children. Per-pupil spending on teacher compensation also grew as a result of more intensive staffing—in particular, the hiring of more resource- and subject-specialist teachers.[5]

In their study of the spending in New York state school districts, Hamilton Lankford and James Wykoff found that the $5 billion increase in expenditures (a 46 percent increase in real per-pupil spending) over the 1980–1992 period was primarily spent on teachers and on disabled students. These researchers suggest that the implementation of PL 94-142 and associated state requirements led school districts to "substantially increase their spending on special education students" at the expense of nondisabled students.[6]

Certainly the needs of special education students are great. The emphasis, however, has been on academic achievement for all. We have been much longer on rhetoric, usually referred to as "standards," than on imaginative and effective interventions to assist teachers in helping ordinary children learn more. New funds go to special education; new demands for academic achievement encompass all students.

These rapid changes in priorities for schools have left teachers and administrators gasping. They have occurred at a time when school personnel, especially teachers, are much older and more experienced than was the case earlier in this century. For example, at the turn of the century the median age of American teachers was twenty-six; hence at least half the teachers were being prepared for teaching and entering the field at a time when assimilation was the reigning emphasis for schooling.[7] Today the median age for public-school teachers is forty-two, and fully 30 percent have more than twenty years of experience.[8] Many of them were thus prepared for teaching in the waning days of adjustment and the rising period of access; neither period demanded the kinds of skills and attitudes for teachers that achievement does today. Therefore, one of the greatest challenges for those involved with schools today is to assist teachers who were trained for one set of school emphases in becoming effective with an altogether different set. For example, I entered teaching the year the Progressive Education Association disbanded (1955), and no one expected me then to teach American history according to today's national standards. How does one mobilize a tenured teaching force prepared to do one kind of teaching to become capable of and willing to do something quite different? That is the mystery of professional development.

In "A Revolution in One Classroom: The Case of Mrs. Oublier," David K. Cohen shows the contrast between one teacher's (Mrs. O's) perception of how a mathematics workshop focused on the understanding of mathematical ideas has changed her teaching, and that teacher's actual practice. Cohen's observations of Mrs.

O's classroom reveal that while the social organization of her class, the teaching materials she uses, and the lessons are all new, Mrs. O's pedagogy has essentially remained unchanged.[9] Schools abound with Mrs. Os, teachers who seek to incorporate new and presumably more effective teaching techniques but who for a variety of reasons fail to do so.

Second, we want our teenagers to get high-school diplomas, but we are deeply ambivalent about what the content of their high-school experience should be. America has led the world in the twentieth century by providing broad and nearly universal secondary education, and for most of the century our means of achieving such widespread access has been to modify the curriculum of the high school, reserving rigorous academic courses for a relatively small minority of our youngsters. The percentage of teenagers who graduated from high school (or in more recent years gained high-school equivalency through a GED) increased dramatically from less than 10 percent in the early years of this century to approximately 50 percent in the middle years of the century; today, 85 percent of American teenagers receive a high-school diploma or its equivalent.

The first federal aid to schools was the Smith-Hughes Act of 1917, which provided funds for vocational education in the high school as a supplement or an alternative to the traditional classical or college-preparatory curriculum. As the fraction of students continuing in high school increased through the middle years of the century, more and more efforts were launched to alter the high-school curriculum to make it more appealing to students who were neither academically inclined nor intending to attend college. The Life Adjustment curriculum at the end of the progressive education movement captured the emphasis on staying in school but not necessarily learning much. In short, the twentieth-century American strategy has been to keep children in school by changing the curriculum while holding the pedagogy constant. A more effective approach undoubtedly would have been to modify the pedagogy, so that it attracted learners, while holding constant the curriculum that society believed all children should learn. But the intent was to prevent dropouts, not to create learners. Furthermore, the effort was successful.

Diplomas were a goal, and thus antidropout programs were important; however, in anti-intellectual America, achievement itself has not been broadly sought for most children until recently. If the approaches to learning academic material were not modified to appeal to a broad variety of students, then inevitably many high-school students, beset with the hormonal and cultural consequences of adolescence, would find high school itself boring. In the absence of a compelling need to be interested in schoolwork, they sought other alternatives. For many—especially for white, middle-class youngsters—the preferred alternative was paid employment.

Working during the school year provides many advantages; mostly, it brings in money, which occasionally is saved but more commonly is used to buy the desirable but nonessential goods that many teenagers crave. Money from working brings these goods without delay. This immediacy appeals to many teenagers whose capacity for delayed gratification—advocated by teachers as a benefit of education—is limited. "Greed, not need" explains the high rate of American youth employment during the school year, according to Ellen Greenberger and Lawrence Steinberg, who argue:

> It is difficult to say whether the increased consumer spending of young people preceded or followed their increased participation in the part-time labor force, but the two forces obviously fed one another. As more teenagers developed expensive tastes and a hunger for luxury goods, they found it necessary to go to work; and as more youngsters entered the labor force and began earning money that they could spend as they wished, more money was spent on developing and expanding the youth market.[10]

Many parents also support their adolescents' employment during the school year. Exhausted parents often take comfort in the security of knowing that their sixteen-year-old is under supervision while flipping hamburgers—and is happy with the money. They take further consolation in the hope that he or she is developing a good work ethic, and some are relieved that they do not have to supervise their child at home, insisting that the homework get done. All, undoubtedly, are thankful that their children are not driving around looking for opportunities to use drugs or alcohol.

Estimates vary, but approximately half of high-school students work during the school year with estimates for white students being significantly higher than for blacks and somewhat higher than for Hispanics. A 1990 New Hampshire survey found that 70 percent of all teenagers held jobs and that more than 84 percent in grades 10 through 12 worked; 45 percent worked more than twenty hours per week during the school year.[11]

These figures are in sharp contrast to the national rate of black teenage unemployment—32 percent of those who are seeking jobs.[12] As Richard Freeman and Harry J. Holzer have observed, "Young blacks have made advances in both occupation and education. Yet their employment problem has worsened, reaching levels that can only be described as catastrophic. . . . In many respects, the urban unemployment characteristic of Third World countries appears to have taken root among black youths in the United States."[13]

Nearly everyone would agree that working a little (less than ten hours per week during the school year) is not harmful, and possibly beneficial. Similar concurrence exists at the other end of the spectrum—working more than twenty hours per week during the school year is detrimental. For those students, grades suffer, less rigorous curricula are pursued, and often health is impaired because of insufficient sleep or inadequate exercise. Less easy to measure is the impact of a short-term job with its immediate rewards compared to the patience required to take demanding courses and work hard enough to do well in them. Although immediate transition to the work force may be easier for someone who has worked extensively during high school, upward mobility is more likely for someone with a strong high-school and college record; calculus is more valuable in terms of discipline than a perfect attendance record at McDonald's.

A recent comparison of youth in Minneapolis and Sendai, Japan, reveals considerable differences in use of time and money. Nearly three-quarters of the Americans work, while only one-fifth of the Japanese do. Average weekly income (from both job and parents) was $205 for the American youths and $86 for the

Japanese; nearly all of the Japanese youths' income came from parents, while only half of American youths' income came from their parents. Other findings included that Japanese watch more weekly television (16.7 hours) than Americans (12 hours) and that fewer Japanese reported experiencing stress each week (43.4 percent) than Americans (71.2 percent).[14]

The United States is atypical in its pattern of youth employment while enrolled in school. Beatrice Reubens, John Harrison, and Kalman Rupp reported in 1981 that almost 70 percent of all sixteen- and seventeen-year-old students were in the labor force during the 1978–79 school year in the United States, compared to 37 percent in Canada, 20 percent in Sweden, and less than 2 percent in Japan.[15] We tend to prolong adolescence while providing the conveniences (but not responsibilities) of adulthood. By encouraging part-time work during the school year, both in high school and for many college students, we delay the time that young people need to assume full obligations of adult life.

Young people often have money for luxuries because they continue to live at home and enjoy a parental standard of living that would not be available to them if they were dependent upon themselves for support.[16] Researchers who examined young-adult living patterns found that between 1977 and 1986 increasing proportions of high-school seniors reported that "living in luxury" was important to them, and that parents tolerate coresidence with their adult children if they did not seek daily funds from their parents.[17] In short, a young person can continue the pattern of high-school employment and using income for frills while parents supply the necessities—as long as the child does not seek direct support for parents for day-to-day expenses. College, too, becomes a much longer process, with many adult students working part time and studying part time, further blurring the boundary between adolescence and full-fledged adulthood. Over half the undergraduates today are over twenty-one years of age—formerly the typical age of graduation from college.

The ambivalence we exhibit about the content of our adolescents' educational experiences is also illustrated by our commitment to high-school athletics. H. G. Bissinger captures the intensity of community enthusiasm for winning high-school football teams in *Friday Night Lights*, revealing adults much more concerned about the teams' prowess than about children's learning.[18] In the face of these pressures brought by students, their parents, and the local community, schools traditionally have relaxed academic demands in order to accommodate student employment and facilitate athletic eligibility. No wonder we are ambivalent about our adolescents' educational experiences.

Finally, how important is school itself in children's education? Scholars ranging from the late James S. Coleman and Lawrence A. Cremin to Christopher Jencks have quite properly reminded us of the limited role that schools play in children's education. Analytically there is no doubt that these writers are correct in identifying families, communities, religious institutions, television, and (now) electronic devices as cumulatively much more important than school alone in the education of the young.

One profound irony of those of us who understand this analytic contribution, and who have the wealth necessary, is that we make enormous efforts to get our children into the best possible schools. As parents, we often select our place of residence based in large part on its proximity to good schools for our children. Although we know that we will supplement our children's education in many important ways beyond what the school provides for them, we still believe that the school itself is a crucial educational intervention for our children. We want schools with good teachers, often defined as ones who have had sound undergraduate academic instruction and who are effective in reaching their students. We want principals with good judgment and effective administrative skills. We want facilities, buildings, libraries, athletic equipment, and computers that will attract and challenge our children. Most of all, we want other students whose families share our educational values. In short, those of us who best understand the limited role schools play in education want the very best schools for our children, and generally we get them.

The irony rests on the fact that while most families want the best for their children, many are not aware of the quality of their schools. As Richard Murnane and Frank Levy point out in *Teaching the New Basic Skills*, the families of the Zavala School in Austin, Texas, also recognized that their children needed good schools.[19] The average annual family income at Zavala is $12,000, and since many families speak only Spanish, the school was the way for their children to learn English. These families saw that success in school was their children's best hope to avoid the poverty of their parents. Imagine the shock of the families when they learned that their children, who had been receiving mostly As and Bs on their report cards, scored in the bottom quartile on the Texas achievement tests. Here was a school that was fooling its clients by pretending that they were doing well, when in fact the teachers were not demanding rigorous work from them. Murnane and Levy recount the pain of the parents when they realized that their children were being cheated by the very institution they believed was their children's best hope for having a better future; they also report how the parents, the new principal of the school, and newly committed teachers, plus some outside advocates, gradually and with difficulty turned the school around and truly improved the achievement of the children.

In short, schools are more important for the children of the poor than they are for the children of the affluent. While prosperous families arrange for a variety of beneficial educational activities for their offspring, school is often the only constructive educational experience that children living in poverty may have. It is thus an extraordinary tragedy that the worst schools—whether in terms of faculty and administrative skills or per-pupil expenditures—serve the children who most need excellent schools, the children of the poor, while the best ones serve the children who have the most educational alternatives, the children of well-educated and prosperous families.

These three dilemmas—the changing central purpose of schooling in America, the ambivalence Americans exhibit about their adolescents' educational experiences, and the contradictory commitments we hold about the relative importance of schooling to education—tax our imaginations to understand and our will to resolve. What is most impressive, however, is the historic capacity of Americans to adapt their educational institutions, albeit slowly, to meet and fulfill the shifting expectations placed upon them. Again, we have the opportunity to demonstrate this ability.

ENDNOTES

[1] I am deeply indebted to Yves Duhaldeborde for assistance in preparation of this article.

[2] Personal communication from Illinois Assistant Superintendent of Education Richard Laine, 14 May 1997. The Illinois median expenditure of all operating expenditure per pupil, including all federal, state, and local funds on an average daily attendance figure based on a nine-month school year, was $4,688. The highest expenditure was in Rondant in northern Lake County, and the lowest was in St. Rose in central Clinton County.

[3] This argument is presented in more detail in Patricia Albjerg Graham, "Assimilation, Adjustment, and Access: An Antiquarian View of American Education," in Diane Ravitch and Maris A. Vinovskis, eds., *Learning from the Past* (Baltimore: Johns Hopkins University Press, 1995), 3–24.

[4] *Education for All Handicapped Act*, Public Law 94-142, 94th Congress (29 November 1975): "An Act to amend the Education of the Handicapped Act to provide educational assistance to all handicapped children, and for other purposes."

[5] Richard Rothstein and Karen Hawley Miles, "Where's the Money Gone? Changes in the Level and Composition of Education Spending" (Washington, D.C.: Economic Policy Institute, 1995).

[6] Hamilton Lankford and James Wyckoff (1995). "Where Has the Money Gone?" *Educational Evaluation and Policy Analysis* 17 (2) (Summer 1995): 195–218.

[7] John Rury, "Gender, Salaries, and Career: American Teachers, 1900–1910," *Issues in Education* IV (3): 215–235

[8] 1993–94 data. See Thomas D. Snyder, *Digest of Education Statistics 1996* (NCES 96-133) (Washington, D.C.: National Center for Education Statistics, U.S. Department of Education, 1996).

[9] David K. Cohen, "A Revolution in One Classroom: The Case of Mrs. Oublier," *Educational Evaluation and Policy Analysis* 12 (3) (Fall 1990): 327–345.

[10] E. Greenberg and L. Steinberg, *When Teenagers Work* (New York: Basic Books, 1986).

[11] Bruce D. Butterfield, "Children at Work: Long Hours, Late Nights, Low Grades," *Boston Globe*, 24 April 1990.

[12] Carol Gordon Carlson, "Beyond High School: The Transition to Work," *Focus* 25 (1990): 4; "Teenagers Who Work: The Lessons of After-School Employment," *Harvard Educational Letter* 2 (5) (September 1986): 1–3.

[13] Richard B. Freeman and Harry J. Holzer, "The Black Youth Employment Crisis: Summary of Findings," in Richard B. Freeman and Harry J. Holzer, eds., *The Black Youth Employment Crisis* (Chicago: University of Chicago Press, 1986).

[14] Andrew J. Fuligni and Harold W. Stevenson, "Time Use and Mathematics Achievement among American, Chinese, and Japanese High School Students," *Child Development* 66 (1995): 830–842.

[15] Beatrice G. Reubens, John A. C. Harrison, and Kalman Rupp, *The Youth Labor Force 1945–1995: A Cross-National Analysis* (Totowa, N.J.: Allanheld, Osmun, and Co., 1981).

[16] B. Hartung and K. Sweeney, "Why Adult Children Return Home," *Social Science Journal* 28 (1991): 467–480.

[17] L. White, "Coresidence and Leaving Home: Young Adults and Their Parents," *Annual Review of Sociology* 20 (1) (1994): 81–102; F. K. Goldsheider and J. Davanzo, "Living Arrangements and the Transition to Adulthood," *Demography* 22 (4) (1985): 545–563.

[18] H. G. Bissinger, *Friday Night Lights: A Town, a Team, and a Dream* (Reading, Mass.: Addison-Wesley, 1990).

[19] Richard J. Murnane and Frank Levy, *Teaching the New Basic Skills* (New York: Free Press, 1996).

Subject Index